Beyond the Skin

Beyond the Skin: The Boundaries between Bodies and Technologies in an Unequal World

By

Bianca Maria Pirani

CAMBRIDGE SCHOLARS

PUBLISHING

Beyond the Skin:
The Boundaries between Bodies and Technologies in an Unequal World,
by Bianca Maria Pirani

This book first published 2014

Cambridge Scholars Publishing

12 Back Chapman Street, Newcastle upon Tyne, NE6 2XX, UK

British Library Cataloguing in Publication Data
A catalogue record for this book is available from the British Library

ISBN (10): 1-4438-6131-6, ISBN (13): 978-1-4438-6131-1

Dedicated to the Memory of Wilde Bettini

Exploring the archaic is the key to inventing the new

TABLE OF CONTENTS

Part II: Bodies at the Mirror of Social Network

LIST OF FIGURES

LIST OF TABLES

PREFACE

ISHWAR MODI

BEYOND THE SKIN: The Boundaries between Bodies and Technologies in an Unequal World by one of the most eminent sociologist on Body, Bianca Maria Pirani, the founder President of the Research Committee on The Body in the Social Sciences of the International Sociological Association (ISA RC 54) is an exceptionally brilliant disposition and examination of the body and its reflections in social behaviour particularly in the context of new emerging technologies.

What this book highlights is the advance transformation of persons into spectators. But this is only one side of the story. The other side of the story is that the *Homo sapiens* have always been and still remain performers besides being spectators. In their role both as performers and the spectators there is lot of emphasis on the skin as well, skin which has to be shining and glittering to make the body as attractive as possible. This is precisely what the modern mass media: films, TV and the internet, and the print world, is churning out and performing day in and day out. The predominant focus of the huge advertising industry is centralized on the concept of beautifying the body. More than anything else, body has become the ultimate focus of attention, more so in the Western world.

What is happening in the West is being emulated all over the world almost irrespective of the levels of development of the various societies and countries. It is another matter that to what extent body is considered central in the thought and action of these societies. It is also reflected in the fact that while there had been so much of discussion and debate on body in its various forms and ramifications in the Western intellectual tradition, body has not received much focus and attention in the oriental intellectual and religious traditions. As against body, it is the soul that has found precedence in quite a few intellectual and religious traditions. Hinduism is the most prominent example of this thinking and thought. It is not body but the soul that matters most for a Hindu. Body for them is only a medium of performing their ordained duties of *Dharma* (righteousness), *Artha* (prosperity), *Kāma* (desire), and *Moksha* (spiritual liberation). It is Moksha or the salvation of soul which is the ultimate objective.

However, the scenario has drastically changed practically everywhere in the contemporary world. Body seems to have gained precedence over soul in spite of a large number of people still thinking that it is not possible to conceive of a soul without its body, and that of a body without its soul. Going much further and beyond these considerations, this book hypothesizes that the body is a spatial-temporal *vector of socially distributed processes which permit the acquisition, storage and manipulation of information through bonds with humans and non-humans, structured in social networks.* As such it is no exaggeration when the author points out that in fact, the 20th century has theoretically invented the body. Body has been theorized in different ways depending on disciplines, ideologies, experiences. The author has rightly pointed out that "thinking about the body implies thinking about who we are, what we are doing and how we are acting".

This book focuses on the boundary between bodies and technology, whose social-technical construction, starting from the *perspectiva artificialis* of the Renaissance, has created the body/mind dualism that characterizes the history of modern culture. In focusing on this variable boundary the author has highlighted the differences that distinguish modern society from pre-modern society, and at the same time investigate the destiny of emotion, memory and unpredictability, which are the constituent values of intelligence. It can hardly be overemphasized that the intelligent infrastructure constitute the next horizon of computer technology.

Bianca Maria Pirani has very competently and brilliantly dealt and unraveled the most crucial unresolved questions within the contemporary mobile transformation through the notion of *the body in action in context: the sensory inscribed body* within the spatial interaction mediated through new socially oriented applications. The author has highlighted the concept of bodies acting in context as the main empirical field which allows sociological investigation to test in *loco* the most important neuroscientific discovery of the last two decades—the existence of mirror neurons. She has rightly pointed out that there is a great need to revise and enlarge our visions of the human condition, to be closer to reality, not only for others, but above all for ourselves. The pericorporeal space revealed by current mobile spaces and experimentally demonstrated by the discovery of mirror neurons, invites cultural forces, and sociology in particular, to a reconsideration of the function of the body in the sphere of its own temporal and spatial systems.

The concluding part of her book: What Homo sapiens teaches the mobile society of the 21st century, is the most original contribution of the author which will motivate many other scholars interested in the field of social sciences / sociology of The Body to undertake researches in this important area of contemporary importance since in her own words "Despite a growing number of reflections about the body, there currently exists no complete works based on the sense and associative function of corporeality as an "entity" located within space and time, particularly in the dominion of social sciences, and more specifically in the field of sociology."

Prof. Ishwar Modi
Director, India International Institute of Social Sciences, India
Member, International Sociological Association Executive Committee

FOREWORD

ROBERTO CIPRIANI

Bianca Maria Pirani, founder of Research Committee 54 ("The Body in the Social Sciences") of the International Sociological Association and for many years its president, attempts to offer the sociological community a sort of *summa* of scientific knowledge about the body and its reflections on social behaviour. This is indeed a remarkable undertaking, the result of long years of study on the subject and hermeneutic experiences of the nature and influence of the somatic in the daily behaviours of social actors. The approach taken by this vast study is multidisciplinary. Here, sociology continually goes beyond its academic boundaries to pit itself, as equal, against the so-called hard sciences so frequently present in the author's reflections. She has chosen a four-fold development, almost an echo of the "four noble virtues" of Buddhism (pain, its origin, its extinction and the way to its extinction), a philosophy and religion that has made the body its *raison-d'être.* These are: the origins of *Homo sapiens*; the rhythmic dimensions of the body; the theories of body in the social sciences; and the dynamics of the body from Renaissance Italy to what is emblematically defined as "electronic autism", indicating a human social reality increasingly accompanied by constant but at the same time continually changing "prostheses" (from the mobile phone to the smartphone, from the lap-top to the tablet, and so on into the future, which will certainly not fail to find other solutions, all having, however, substantial synergy with our bodily potentials).

This is a work that digs deep, often revealing what lies under the skin of the topic under investigation. In it, Bianca Maria Pirani offers lucid readings of reality. Far from being vague pronouncements, what is presented here represents conclusions reached only after long reflection and continual modification on the author's part.

The keystone of the first part of the development is what holds together the first two parts, where, starting from the profile of *Homo sapiens,* the Working Memory is gradually delineated. Nor is it by chance that the very real metaphor of the dance embodies – and there couldn't be a word more apt – the features of a "rhythmic body" that is both a form of self-expression free from social trammels and limitations and an exercise of memory, of respect for times, of convergence between gesture and sound (real or only imagined), as we admire in the famous age-old Addaura Cave, in Sicily, which has often moved and impressed the author, inspiring suggestions for future analyses. The symbiosis between mask and dance reconfirms what has already been said, and leads the author to move between reality and so-called, more or less theatrical, fiction. (And is it not true that theatre is part of life? And that our own agency is the producer of plays already given and/or promised?)

In a context of such detailed and at times inextricable development, Bianca Maria offers a thread that guides the reader through the labyrinth of theoretical formulations and empirical findings to find the path of final explanation-interpretation. To this end, the third part gives full treatment to the importance held by the theme of body in all the social sciences. And once again, at the point of arrival, we find the “rhythmic body”.

Probably the most original contribution is found in the concluding section, the fourth, where input of various natures and origins comes together. This input derives from “modern culture” and includes reference to the very recent concept of mirror neurons, a further proof of the necessary, and therefore undivided and undividable, interface between body and social environment, individual and society.

Finally, the author attempts to offer a look forward to the future, with *Homo sapiens* becoming Working Memory to teach coming generations the paths of action.

INTRODUCTION

"We are our bodies". "We have our bodies". "We make our bodies".[1] This "three headed" axiom has turned the body into the "parasite" of modern culture. This is the image that the Canadian director Cronenberg skilfully used in the film *Shivers* to illustrate the body as the site of the most unsettling transformations and intense effects. This "biological horror" – exemplified by bodily illnesses, mutations, strange creatures, violent telepathy, video hallucinations, drug addiction and perverse sexuality, the focus of Cronenberg's film – testifies to the status of the body, in the literary and cultured western world, as a "space sanctioned for the expression of internal conflicts" (Lemma, 2001: 28). A space that unwinds, mockingly, "beyond the skin" of human beings, animals, things, and emerges as a sort of multimedia icon, in "synthetic worlds", currently presented in every possible format by the so-called "convergence culture" (Jenkins, 2007).

What our bodies do to us is just as critical as what we do to our bodies: our bodies age us, and eventually kill us. Bodies are strange, not only in their plurality and the elusiveness of their meanings and interpretations, but also because we theorize the body in different ways depending on our disciplines, ideologies, experiences, and also on our own bodies. Thinking about the body implies thinking about who we are, what we do and how we act.

Despite the growing amount of reflection about the body, there currently exists—particularly in the dominion of the social sciences, and more specifically in the field of sociology—no comprehensive work based on the sense and associative function of corporeality as an "entity" located within space and time. It is not a question of reconstructing the history of

[1] Within this axiom the body is presented by the diverse sociologies of the body as "subject", "object" and "project". Or, using different terminologies, as "self", "other" and "action". The first assumption, "body as subject", mainly refers to the theories of Sartre and Goffman; the second, which considers the body as "object-machine", refers to Descartes' formulation and to its fundamental function in bio-medicine; the third, "body as project", specifically refers to the intersection of the body with gender, race, social class, and in a different way is also present in the works of Giddens, Bourdieu, Bromberg, Foucault and Butler.

the body and its conceptualization, nor of revisiting the paradigms that have been used to think about the body. Instead, it is a question of focusing on current paradigms for identifying corporeality as the total of sensoriomotor abilities that permit the body to interact successfully with the environment that surrounds it. The determination of the boundary between the social *sensorium*[2] and today's technology is the fundamental and as yet unsolved methodological problem emerging from the current "*spatial turn*"[3], a development coming out of the reorientation of the classic digital paradigm (a concept that is centred on the opposite idea of the disappearance of space, the end of geography, the reduction of mobility, and the annulment of embodiment in digital communication on the web). In the evolution of contemporary mobile communications, the concepts of space/code (integrated with the communicative and social revolution of the amplification and portability of social networks) represent the foundation of the prospects for new mobility paradigms which technological developments, and more specifically the ecologies of wireless communications, are producing and articulating in post-PC society.

The impact of physical mobility on the logics of contemporary culture is such that sociological research is reorienting its analytic paradigms to make mobility the fundamental paradigm of our daily lives. However, this return to the relevance of spatiality, in both the physical and the offline worlds, is not a return to space in the sense we have understood up to now—a mere container of individuals, objects and events. As Dourish and Bell (2011) have shown, in light of the processes of digitalization of the omnipresent web, the re-emergence of spatiality in diverse disciplinary dominions (geography, sociology, architecture, economy, art, and urban informatics, to name some), is configured as "transduced space"[4], or in other words as space constructed and activated continuously and socially by the networks and new digital realities emerging at the intersection of mobility, geo-sociality and augmented reality. Within this *scenario in*

[2] In English, the term "*sensorium*" signifies the whole of the sensory functions that in a given moment
connect the human body to the context in which it moves and acts. (see Ragazzini, G. (2005), *Dizionario il*
Ragazzini, Bologna: Zanichelli: 972).
[3] "*Spatial Turn*", in the words of Warf, Arias (2009).
[4] The theory of *transduced space* is taken from Dodge, Kittchin (2011) in an essay in which the authors refer to their previous works as well as to the work of A. Mackenzie (2002, 2006) on the social dimensions of software and programming codes.

movement there emerges the concrete problem of delimiting the space of action as an *elementaryintegrated unit,* defined by the interaction between the sensorimotor activity of the experience of the body in a given situation and the mobile connection produced by a specific technology.

This book uses the Freudian concept of "drives"[5] and the insight of Lévi-Strauss, who holds that the most important and effective of the inherited instincts in our biological patrimony consists of a "mass of unconscious rules"[6]. It considers the somatic dimension as the biological basis of any performance situation that does not require conscious attention to be brought to mind, generally in the form of habits, perceptive strategies and abilities acquired through experience. Based also on some of the more recent acquisitions in neurosciences (Damasio, 2003; Kandel, 2005; Gazzaniga, 2008; Goldberg, 2009), we propose the somatic dimension as the foundation of the physical bond that connects the body to the surrounding context, anchoring its associative pathways to the places where we live.

The somatic-sensitive areas of the brain (Damasio, 2004: 314) construct the dynamic maps that the boundaries of the body configure with the environment through pathways of neural activity coordinated over time. Without this mechanism, we would not be able to localize our interactions with the environment, nor would we be able to use the wealth of knowledge acquired by the body when touching an object, looking at a panorama, or moving through space along a trajectory that it describes through its movement. Most interactions between the body and the environment take place within a *spatial and temporal limit* generated by the combination of the boundaries of the body and the practices of our culture, whichever of our senses is used (the sense of touch or any other). This combination is a fundamental ability for our mental world. Without this skill, every object and problem would be an *ex novo* encounter, and we would not be able to use any of our previous experiences to deal with them. As the works of Herbert Simon (1996) have demonstrated, the recognition mechanism is the main one used for problem solving in present experience. This mechanism is morphologically identifiable as the

[5] Freud (1915: 17) had identified in the concept of "drives" the "*boundary* between the psychic and the somatic" (cursive added).

[6] In a text for UNESCO published in the 1980s, Lévi–Strauss (1984:42) strongly supported the idea that "between the instincts inherited by our biological patrimony and the rules inspired by reason, the mass of unconscious rules remains the most important and most effective, from the moment when reason itself, as understood by Durkheim and Mauss, is a product rather than a cause of cultural evolution".

working memory, which organizes behaviour with respect to the temporal dimension and controls the execution of the operations the body enacts to achieve its objective in the environment.

The capacity to recognize is not exclusive to the human race. It is shared by every other species with the capacity to learn. What differentiates human beings is their capacity to transmit inventories of recognition from individual to individual and from generation to generation through cultural practices.Indeed, the architecture of the organism is neither innate nor acquired, but is instead the result of continuous interaction between endogenous genetic factors and exogenous environmental factors. The construction of this architecture distinguishes animal existence from human existence. The transformational evolution from animal to human existence entails a radical change in the nature of transactions between the body and its environment: animals always build the same dens, and as far as we know they have always built them this way (Ingold, 2001: 1116), while human beings build different types of houses. Although these types have existed for a long time, it is a historical fact that they have undergone significant changes over the course of time.

Based on these considerations, now fully acquired by neuroscience, this book hypothesizes that the body is a *spatial-temporal vector of socially distributed processes which permit the acquisition, storage and manipulation of information through bonds with humans and non-humans, structured in social networks.* Every process of human formation regards a specific individual who relates to contextual variability *with* and *through* his or her body, which acts within a *determined spatial reality* characterized by a specific symbolic system of reference (contents, beliefs, culture and language). The comparative analysis of the characteristics of corporeal-kinetic intelligence sheds light on its capacity to shape itself in close relation to spatial intelligence, which centres on individuals' ability to transform objects in a specific environment and orient themselves in a world of objects in space. It is this capacity to orient ourselves in space that allows us to establish our position in the environment, to estimate the distance between external stimuli, and to move. It is this very complexity that has up to now prevented a systematic analysis of the fundamental dynamic relationship between corporeity and the surrounding space, especially as regards the processes of abstraction and emptying out that the intellectualization of experience has caused in modern culture.

The 20th century invented the body theoretically. This invention came out of psychoanalysis, when Freud, observing the exhibition of bodies displayed by Charcot at the Salpetrière Hospital, identified conversion hysteria and understood that this would be the starting point in the development of his studies: the unconscious speaks through the body. That first step was decisive, because it led to the idea of somatisation and the consideration of body image in the formation of the individual—what was to become the "skin self"[7]. The next step, made by Husserl, was to focus on the body as the "cradle" of all meaning. Husserl's influence was particularly felt in France, where it led to Merleau-Ponty's conception of the body (1945: 97) as "embodied consciousness" and "cornerstone of the world". The third step in this discovery of the body was the result of Marcel Mauss's surprise when he noticed that the British infantry in World War I marched to a different step from the French. This notion of "body technique" has profoundly influenced all historical, anthropological and sociological reflection on the question of the body.

However, despite Mauss's enlightening intuition, more often plundered than sufficiently understood, the body was relegated to the subconscious, stowed away inside the "subject" and included as the "own body" among the social icons of culture. It was as if knowing, in the most general sense, actually signified "representing", according to the philosophical view that contrasts the immediate data of consciousness to the principal elements of representation, a view that sees culture as a social substitute for instinctual automatism. The crucial turning point of modern European history is found just here: only after the transformation of the world into a series of limited but open models (in that they continually refer to something that is *absent* because it is *external*), does the *space for representation* guaranteed by paper become possible. To quote Lefebvre (1974: 59), space is that which "the imagination attempts to modify and occupy". While in pre-modern society space constituted the *anchor of time*, in modern society space flattens time by standardizing social organization. According to Giddens (1990b), this change coincides with the expansion of modernity.

Thus, this book focuses on the *boundary between bodies and technology*, whose social-technical construction, starting from the *perspectiva artificialis* of the Renaissance, has created the body/mind dualism that characterizes the history of modern culture. In focusing on this variable boundary we highlight the differences that distinguish modern society from pre-modern society, and at the same time investigate the destiny of emotion, memory and unpredictability, which are the constituent values of intelligence.

---

[7] Cf. Anzieu D. (1985), *Le Moi-Peau* (1985), Paris, Dunod; Bick E., "The Experience of the Skin in Early Object Relations", *International Journal of Psychoanalysis*, 1968, 49: 558–566.

Contemporary information technologies come to grips with this dimension through the production of *lightweight, high-density materials* (Maldonado, 1992: 80). The physiological, sensorial and cognitive mechanisms that are activated in the relational context through current information technologies are the same ones found in pre-modern relations between man and technology. The Swedish neuroscientist Torkel Klingberg (2008: 10-11) has proved that today's brain capacity is identical to that of the oldest anatomically modern *Homo sapiens*. What has changed in *Homo sapiens* over the long course of time from the invention of the first stone tools to today's information technologies?, What is dramatically clear, starting from the perspective of the Renaissance, is the exclusion of *biological, material and corporeal "situativity"* in the construction of sensorial experience that humans create in the environment. From the chipped flint stones of the Upper Paleolithic to nuclear power plants, the technical resources of human ingenuity have responded to the primary biological need to dominate the environment, which is inseparable from human nature. While the biological history of living forms has always been determined by ecological interactions with the environment, humans have created their symbolic and cultural universes of meaning, using them as a *diaphragm* between themselves and nature, and imposing them on themselves and on nature as a model. In the experiment of worldwide significance that technology has started to perform on humanity, our innermost root is coming to light: the "body in action in context", the foundation which the *rational animal* has always stood upon, and which in our depths unites the possibility of reason with the animal nature of the living thing. Intelligent infrastructures make up the next horizon of computer technology.

We are entering the phase of microelectronic technology in which computers will interact in a totally new way with the physical world. Today we have to discover that, even though the world may be restricted in terms of physical access, its horizons continue to expand through the contextual assemblages that create mobile boundaries between sensory experience and technologies. The technological environment that surrounds the body profoundly influences its ways of being, doing, and bonding with the surrounding bodies of living beings, things and places. In considering technology as the matrix of the agent/context relationship, the theme of corporeity is considered from a bio-dynamic perspective, aimed at investigating the adaptive individual/environment relationships as an inseparable mind/body whole, encountering contextual variability.Today we know that the motor system, far from being a mere muscle- control system or simple executor of encoded commands, is able to perform cognitive functions that for a long time were erroneously considered to be the prerogative of psychological processes and neural mechanisms of a purely associative kind. For neuroscientists it is now a confirmed fact that human beings automatically and unconsciously organize bodily experience in spatial units, whose boundaries are established and determined by changes in place and time and by the interactive elements that determine this experience. These processes constitute the *implicit* or *procedural memory,* a form of memory that is not recalled at a conscious level. Already conceptualized in French philosophy (Maine de Biran, Ravaisson, Bergson), this memory governs not only simple perceptual and motor abilities but also Margot Fonteyn's pirouettes and Wynton Marsalis' trumpet technique, as well as cyclists' leg movements. Descartes himself observed that the lute player has hand memory when playing his instrument. Procedural memory constitutes the biological foundation of all performative situations. It is often automatic and is directly accessed through execution, without any conscious effort or awareness of reliance upon memory. In its contextualized *performances*, implicit memory guides us through consolidated routines, abilities and habits acquired through experience.If explicit knowledge can be communicated through language with a high degree of awareness, implicit conscience has a character that pertains fundamentally to the somatic, emotional and behavioural dimension, referring to the "here and now" of the "present" (Stern: 2004).

The "body in action in context" is the "procedural" lesson that primitive *Homo sapiens* has transmitted, in the findings documented by paleoethnology, archaeology and art history, to the embodied spaces of the digital era. The secret of memory is written in this humble and pragmatic lesson. The capacity of the contemporary brain is identical to that of anatomically modern *Homo sapiens.* What changes is the volume and complexity of the information that constitutes the "technological environment". Technology is the place of relation/mediation among actors, their social networks and their cultures. It is neither neutral nor inert. Throughout history, when a technology once conceived of as fixed became mobile, a cultural change has accompanied this transformation. When, for example, the technology of writing was transferred from stone inscriptions to writing on papyrus or paper, the world changed.A similar cultural shift also occurred at the moment in which computing technologies became mobile.

As Intel announced in the year 2000, "Computing, not the computer, will characterize the information era". This is the key point of the current cultural transformation: it consists of an *activity*, as opposed to a playful pact "on this side of politics and culture" (Susca, in Jenkins, 2007: 364), within which the maps of popular cultureare traced[8]. Castells (2009: 75) defines this unprecedented form of communication as "electronic autism", with reference to the social spaces of simulated reality, which combine the sociability and desire to experiment of grassroots convergence with role games. "Electronic autism" is the concrete proof of the risk of wanting to investigate and manage new phenomena with cognitive tools and methodology inappropriate for the analytical challenges of our day and of the near future. In classical web analytics, the dimensions of mobility and spatiality have not been the object of particular attention or in-depth research in recent years. The examination of social networks is, in fact, all too often conducted without inserting the variables of space and time into the analysis, evaluating only the abstract topological dimensions of social graphs. In this framework, the growth of information asymmetry between data actor and data collector is directly proportional to the uncontrollable expansion of manifestations of "electronic autism". To contribute to avoiding this risk, using the notion of *the body in action in context* this book proposes to treat one of the crucial, unresolved problems of contemporary mobile transformation: the sensory-inscribed body within the spatial interaction mediated through new socially oriented applications.

[8] The English term "Popular Culture" indicates the scenario of the New Mediaas a set of cultural practices and features that portray how individuals and societies relate to means of communication. Today, the large majority of cultural products are not for the masses. In fact, we live in a world of infinite niches and sub-genres. (See Jenkins, H. (2006), *Convergence Culture*, New York, New York University Press; Ital. ed.,.Jenkins, H. (2011), *Cultura convergente*, Milano, Apogeo.)

PART I:

THE HAND AND THE TOOL: THE INVENTION OF TECHNOLOGY

CHAPTER ONE

THE HAND AND THE TOOL: THE 'EMERGENCE' OF *HOMO SAPIENS*

1.1 The Heredity of the Social Intelligence of Animals

The emergence of *Homo sapiens* is the unexpected outcome of contingent adaptations. Since the evolution of the biped, the arrival of *Homo sapiens* with his modern modes of behaviour represents the advent of an unprecedented species. Arguments reducing the emergence of *Homo sapiens* to no more than an improvement of the species which preceded it are unsustainable. What is globally defined as 'human faculties' comes under the category of emerging properties that allow a new combination of elements to produce totally unexpected outcomes by chance[1].

The question of our origin lies in the reconstruction of historical, genetic and adaptive modalities that gave rise to the explosion of the human species. Such an 'explosion' was the result of a long, uninterrupted evolutionary process which generated *Homo sapiens* through speciation (the evolution of new species was a continual process): the unexpected outcome of contingent adaptations. To explain this result, a new approach to evolution is necessary. Ingold (2001: 78-79) observed that not only was there a need for an approach that enabled us to explore the variations in, and the selection of, transmitted characteristics from generation to generation, but there was also a needto investigate the dynamics of self-organisation and the potentiality of self-formation. It is the living body that provides the cognitive key which explains the processes and the remarkable combinations that connect the first hominids with the appearance of *Homo sapiens.* It is the living body which includesits relational matrix, the individual/environmental bond, the*incorporation of difference.*

---

[1] Water, the classic example of 'emerging properties' essential to life, possesses characteristics that cannot be separated from its components hydrogen and oxygen, considered as two separate elements.

Konrad Lorenz argued that life could be described as a "process ofknowledge". The biological adaptation of organisms is, in fact, analogous to the acquisition of information relative to the environment. After each experience, new information is stored in the nervous system, the organism interacts with its environment, and a more or less permanent behavioural adaptation - usually called 'learning'- is determined. Only animals behave, or rather are active, in order to carry out certain functions and achieve certain ends. The inanimate world has neither goals nor purpose. Function appears with the first living structures. No metabolism or physiology exists without organic structure, no matter how small or large it might be. Their conformation or assemblage is such that they appear as if they are meant to carry out a particular function. Even the simplest cells have a behavioural embryo: they are repelled by toxic substances and attracted to nutritional matter. Elementary responses to external environmental stimuli can be regarded as *functions*, functional units or behaviour. These multi-cellular organisms coordinate particular cell functions, and above all control body activity, *movement* or a series of coordinated movements known as *actions*. For this to occur there needs to be a nervous system that elaborates stimuli from the external environment. Whatever happens between these two events is called the *mind*. Today, there are a number of scientists involved in the study of the mind and the brain, observing animals of various species. According to Vallortigara (2007: 15):

> they are convinced that the most fundamental principles concerning the function of the mind - all minds, including the human mind - can also be obtained from the study of organisms that are very different from our own.

The basis to an understanding of the processes that occur between the nervous system and receptor organs, between movement and the outside world, is the study of the animal world (Umwelt). 'It is that part of the surrounding world which acts through receptive organs' (von Uexkull, 1934). The sensorial abilities of our species, like all other species, are simply the consequence of natural selection.

Evolution means, above all, progressive differentiation. Living beings change over the course of time as new structures appear that can co-exist with the old for long periods which in turn might undergo further changes. Consider the remarkable variety of plant life, birds and dinosaurs which has been found over the last two hundred years. Evolution brings about a transformation and an increase in variety, added to which, although not always, there is an increase in complexity. Take, for example, the extraordinary refinement attained by organs such as the eye or the ear over hundreds of millions of years. Or, another example, the development of the human body, now capable of activities unknown to our ancestors who millions of years ago were learning how to use fire and break the bones of animals to suck out the marrow for nutrition.

Neither natural selection nor chance, nor the wider-scale processes of selection, can affect physiological or ethological systems in isolation, whether they are of the same species or in the transitional phase from one species to another. Even if the gene constitutes a discrete unit, it cannot function in isolation, either separated from the entire genome or cut off from the immediate surroundings of the cell. The exclusive characteristic of life is that it can reproduce itself, if conditions are suitable, and it can assume an enormous variety of forms, as the world surrounding us testifies. Even a crystal can reproduce itself, each identical to the parent crystal. A bacterium, too, has an identical structure to its parent, but over the generations it can change, revealing a complexity that was not present in the crystal; it continues to reproduce, remaining almost identical to itself, while simultaneously evolving.

In his experiments on song birds, Marler discovered that they do not sing the tune of their own species – it is not innate – but rather, they have the innate ability to recognise their own tunes from those species they live with (Marler, 1970). These birds learn to sing selectively, discriminating and favouring the structures sung by their own species. If they are taught short tunes with some variations, it is possible to identify the features of the song that males of a given species will imitate; these attributes vary from one species to another.

Learning the tune constitutes a very clear example of the close interaction between the cultural transmission process of a melody and the innate predisposition of the other. These discoveries led Marler and Gould to speak of 'learning instincts' (1987).

The song bird is eager to learn to sing, adopting a particular vocal tradition. It would seem that it is also inclined to embellish what it has learned, improvising and inventing. Such a rich potential of behavioural malleability with a possible genetic basis is conditioned by an innate predisposition. No matter what the consequences of this interaction between genes and culture are, every species has its distinctive song domain. In every domain that constitutes a potential cultural space, the range of variation can be huge; the boundaries for the species define a coherent, possibly complex vocal unit, even though they might change over time within their own cultural space. The 'song of the species' is a behavioural umbrella that presents hundreds of variations on a theme. This repertoire of songs forms the nucleus of the system of vocal communication that controls different types of interaction between sexual partners and territorial rivals. The 'learning instinct' concept suggests that one species can manage to use cultural diversity to its advantage, while simultaneously maintaining a given degree of cultural canalisation.

Evolution means, above all, *developing the capacity to interact with the environment*. It is this which determines the success of a species or its extinction. As Hamilton (1964) argued, organisms develop through continuous interaction with the environment, from the embryonic stage to adulthood, whether self-generated or externally stimulated.

Environmental space is at one with *the tactile sensitivity of the living body* and its movement in context. Researchers have found that the cerebral cortex of superior mammals is generated by a network between functions of perception and motility made up of processes that join neural networks to receptive organs and *Umwelt.* Such a space (exclusively associated with interactive events that connect sensorial receptors to environmental elements) is structured by motility sensory integration which supports learning activities and the exploration of active contexts. Through its movements, the animal creates its own territory of action; the combination between the 'pericorporeal' space and the elements can at any given moment constitute its living environment. In this variable space, the animal lives in its own sensorial world and carries out its own territorial activities (Hall, E.T., 1988:15).

In animals we may observe direction, frequency and extension of behavioural changes that depend on the variation of space available. We could never hope to repeat similar experiments with humans. First of all it is possible to accelerate times with animals, since animal generations are relatively short. In forty years one scientist can observe four hundred and forty generations of mice while over the same period of time he or she will have seen only two generations of man. There is also the fact that since animals do not rationalise their own behaviour, they do not hide the regularity of the results: in their natural condition, they respond in a wonderfully coherent way, thus making the study of repeated and virtually identical actions possible. Limiting observations to the way in which animals use space, the observer can understand a quantity of translatable things in terms of human behaviour.

Territoriality can be defined as the characteristic behaviour with which an organism claims its rights to an area, defending it against his own species. The animal psychologist, Hedigher (1995), teaches us that territoriality guarantees the propagation of the species. It regulates its density, providing patterns in which actions are performed; it identifies strategic hiding places and suitable positions for learning or play; it coordinates group activities and maintains unity; it establishes distances for animal communication regarding the whereabouts of food, or in the case of enemy threats, it raises the alarm. Carpenter, a pioneer in the study of monkeys in their natural habitat, listed thirty two functions of territorialism, among which are those that particularly pertained to the protection and evolution of the species (Carpenter, 1958). Territorialism is also a behavioural system related to spatial changes generated minute by minute by the tactile sensitivity of the body in movement.

Animals are born with not only the capacity to perceive and act, but also to learn and use experience gained in the past to improve behaviour in the present. Association and habit are two recognised forms of behaviour in the animal world, and as far as we know, two of the most diffuse. Learning is not automatically acquired, it is comprised of a series of specific mechanisms, the product of evolution. As we have noted above, the variation of the learning 'toolbox' available to living creatures is particularly evident. Song birds know how to learn: they can distinguish the differences between diverse learning mechanisms and diverse species.

There is less diversification between animal cultures than there is among humans. Although there are several differences between groups of animals (the baboons from the savannah, for example, tend to live in large groups, while those from the high plains congregate in smaller numbers), the complex effects on the life of each animal are more limited. In her research, Jane Goodall (1971), the primatologist, illustrated the complexity of the life of the chimpanzee, demonstrating that the construction of interactive events is central to the social intelligence of this species. She carried out an in-depth analysis of social life both in the family and the group, the communicative strategies within and outside the species, and the cognitive structure in finding solutions to problems. The principal contribution of her research into cognitiveethology and animal bioethics (two inseparably correlated disciplines) was the formulation of the concept of 'animal culture'. Animals, too, demonstrate the capacity to create new procedures that are not simply the outcome of a phylogenetic response of species-specific adaptation. On the contrary, such new procedures are the effect of a very real and singular innovation that emerges at an ontogenetic level. These procedures are later transmitted to the social complex, mediating learning mechanisms that enforce group identity. For example, some chimpanzees learned to use small twigs as a kind of hook to extract termites, a favourite food. Such results bear witness to the fact that chimpanzees are capable of constructing their own tools, hunting with complex team strategies and elaborating a very real local and transmittable culture. Small chimpanzees played amongst themselves with an infinite number of games, the most frequent being chasing a playmate around a tree or climbing from branch to branch, catching a playmate or swinging, wrestling, tickling and so on.

These activities serve to develop procedural capacities, useful both in movement and future combat. As Goodall pointed out, the game plays a major role in learning skills. The young chimpanzee quickly identifies from among his peers the strong from the weak, the fast from the slow; those who tend to dominate from those who are more submissive. He also learns that some mothers intervene in defence of their offspring, while others can be intimidated by a show of aggression; some even try bluffing. The success of the young monkey's learning depends on his capacity for confrontation and remembering social events correctly. A large group needs a considerable cognitive capacity. As the group increases in number, so too does the number of relationships to remember. Dunbar (1988) observed that large groups of primates tend to lack stability unless their members maintain regular dyadic relations with their environment. To achieve this, a significant memory capacity and an understanding of complex social relationships are required. Social intelligence has never been adequately defined; for modern human beings it is perhaps difficult to conceive of a symbolic expression to which intelligence might be attributed, a 'language' that preceded the spoken language.

Intelligence is not an island but an archipelago. Vallortigara (2007:65) observes that 'intelligence consists of a set of general abilities common to all species, more than a set of specialised abilities, each one evolving as a solution to a specific problem'.

Thus, a young chimpanzee's social ability is based on a complex series of time/spatial relationships onto which he constructs his mnemonic map.

The sensorial abilities of our species evolved like those of other species: in response to specific needs imposed by the diverse demands of adaptation throughout the course of natural history. According to scientists, the differentiation between humans and chimpanzees began between 6 and 4 million years ago. The evolutionary line of hominids ceased around 5 million years ago. It is a process marked by anatomical and cultural events that date mutations of major significance to 4 million, 1.5 million and 200,000 years ago. Between these millennium 'markers', hominids experienced periods of slower anatomical and cultural changes including fewer variations of an extreme nature.

The appearance of the Australopithecine and the accompanying anatomical and cultural mutations dating back 4 million years affirmed several essential social structures which are still characteristic of the social human of today, such as bonding between couples and social cooperation regarding the provision of food and the care of offspring. But the most striking event that highlights the appearance of the Australopithecine is that he stood erect, on two feet. Besides making adaptations to arboreal life - tree climbing and swinging from branches - the biped Australopithecine underwent numerous anatomical transformations of not only the pelvic position and the length of the leg-bone, but also the feet, the hands, the chest and the respiratory system. As Lovejoy (1980:250) pointed out 'the adoption of the erect posture meant an adaptation of survival strategies such as the sharing of food, procreation rhythms and methods in child rearing'.

The Australopithecine is a biped primate, terricolous, with manual ability and social organisation very similar to humans. On the basis of geological distribution, it is highly likely that the Australopithecines inhabited a large part of the African continent. The best known of these hominids is the *Australopithecine afarensis*; the most comprehensive example, dating back 3.5 million years ago, is exhibit A.L. 288-1, commonly known as Lucy, found at Hadar in Ethiopia.

The period between 4.5 and 1.8 million years ago was characterised by the presence of biped anthropomorphic monkeys in Africa, each with a slightly different corporal and dental structure, revealing particular behavioural patterns and eating habits. Small teeth and larger bodies are distinct differences which help to define the *Homo* type (Wood, Collard, 1999). As the anatomy of the first hominids suggests, they were equipped to inhabit relatively open spaces, probably in more numerous groups than their immediate predecessors, ranging from perhaps eighty to ninety individuals (Dunbar, 1998). Microscopic analysis of implements for wear and tear (Klein, 1989) indicates a more sophisticated utilisation of tools for the cutting of meat, leather, bones, copper and plants. The turning point in human evolution was the appearance of *Homo ergaster,* who livedfrom 1.8 million to 900,000 years ago. His development is totally determined by his evolution as a biped. The so-called 'Nariokotome boy' skeleton, found in Kenya,dates back 1.6 million years. With *Homo ergaster*, body movement becomes the focal point of adaptation: his feet are collocated relatively close together on the ground and his thigh muscles are used to contract hip oscillations, with a broader pelvis, a slanting lower backbone and a cranium in a vertical position. The anatomic adaptations in the upper part of the biped body gave rise to a long step, each leg moving alternately. From the partial bipeds of the Australopithecine to the fully modern bipeds of *Homo ergaster*, the hominids learned to move through the territory, exploring by trial and error, defending themselves from aggression, and elaborating sensorial information in a useful, strategic manner for the individual and the group. In brief, *Homo ergaster* had become a complete interactive body in motion: *Homo sapiens*.

1.2 The Invention of Technology

Man is inseparable from his tools, and evidence of cultural evolution demonstrates this. *Homo habilis* was already using rough stone implements 2.5 million years before *Homo sapiens* appeared on the African savannah. Knowledge of the simultaneous evolution between humans and tool production has caused significant modifications in evolutionary thought. According to early hypotheses, evolution was first biological and only later, cultural. Today, current thought is of the opinion that *tool-making andHomo sapiensevolved together*; the manufacture and use of tools had a primary function in the development of the brain linking modern man to his predecessors, the first humans.

Of all human activity, technology is the only one that does not go back to the starting point. Once a technique has been mastered, knowledge continues to accumulate; improvements are imperceptible, similar to the evolutionary process of living beings. As Leroi-Gourhan (1993:35) observes:

> For one who journeys far and wide over the dusty space of thousands of years, the problem is the tie linking technology to man's brain - knowing whether we are humans because we make tools; knowing whether (as philosophers would have us believe), it is possible to conceive two types of humanity – at first following and then as part of the life of every human, the humanity of *Homo faber* and the humanity of *Homo sapiens*; knowing whether one thinks with the same parts of the brain like *faber* or like *sapiens* or whether the first humans started out with a brain in which the *sapiens* thought had only a very limited number of cells and whether, right from the start, the technical brain functioned faster than the 'cerebral' brain. Or in more orthodox terms, can we ask whether the cortical centres of manual motility and the associated zones in the recognised expansion of the cerebral neo-pallium, did not precede the development of the complex system in which language triggers off the elements of intellectual thought.

Recent developments in the neurosciences demonstrate Leroi Gourhan's grounds for posing such a question. In the development of thought, doing comes before understanding. The mind originates in the body and such deep-seated roots can never be substituted by a mere description. The hypothesis of the 'dynamic nucleus' proposed by the neuroscientists Gerard Edelman and Giulio Tononi (2000:165) regarding conscious experience demonstrates that 'no scientific description or anything similar is equal to the experience of corporal roots.'

Using a tool generates the capacity to repeat an action, adapting it according to the context. This capacity is an *emergence* that implies a kind of correspondence and stabilisation between response and environmental states. Thus, *the exercise of technology constructs the field of interaction between the individual and society in the body, both systems of neurological processes being regulated by experience.*

The development of the technical skills that the body learns when handling objects and tools explains the specificity of human cognition with respect to animal intelligence such as *the creative process that connects* useful *pre-existent elementsmaking new combinations*; a 'tool box' for inventions suitable for every new experience, the 'skilled practice' that all those years ago Marcel Mauss (1936:394) identified as a 'primary instrument' of cultural reasoning.

As Sills (1968:58) emphasised, technology is a creative system generating *knowing how.* Ryle (1949) expressed it as 'know how', the ability to carry out specific procedures based on the recognition of specific stimuli. The basic function of technology is to situate the body of the human being in the environment he inhabits, generating *forms of exchange* that constitute *social space.* The adaptation procedures correspond to the means of intervention on matter, facilitating self-conservation and the transmission of the body's genetic heritage. The human nervous system, and the brain in particular,evolved by solving problems ranging from how to avoid drowning to finding food or a companion. Each new technological invention is *an active manifestation of a new relational product* whose appearance is determined by the unrepeatabilityof matter, events and the participation of human beings. The basic drive is the imperious need to expand, extend, develop and activate the sensorial capacities of the body to its limit so that it may improve and appreciate in value. Again Leroi-Gourhan (1977, 11:329), shows that 'from the sensorial point of view, the most elementary behaviour in animals can be reduced to three levels: nutritional, whereby the functioning of the body's organism to assimilate matter is guaranteed; physical affectivity guaranteeing the genetic survival of the species; and the spatial integration necessary for the above two levels to function'.

These three levels diversify according to the degree of evolution of the species, they compare and contrast variable relationships of diverse instruments of sensorial systems. Above all, the hand directly activates the processes of the chemical-physical exploitation of matter in 'complex operations of pressure-rotation-motion that characterise manipulation'

(Leroi-Gourhan, 1977, 11: 284). It was the privilege of the hand to carry out these first operations, the very archaic and not very specialised human hand. Compare it 'with the lion or the horse with their wonderful paws for grasping and hooves for galloping' (ibid. 1977, 11: 285).

The image refers to the organisation of the cerebral system that distinguishes the human from the monkey when using technical implements. Leroi-Gourhan highlighted the predominance of the hand, activating both coordinated prehensile actions and the preparation of food, as well as actions related to attack and defence. The integration between motility processes and processes of manufacture is achieved with the aid of the technical tool; *the process of adaptation through which the human body intervenes on surroundings only as far as necessary.*

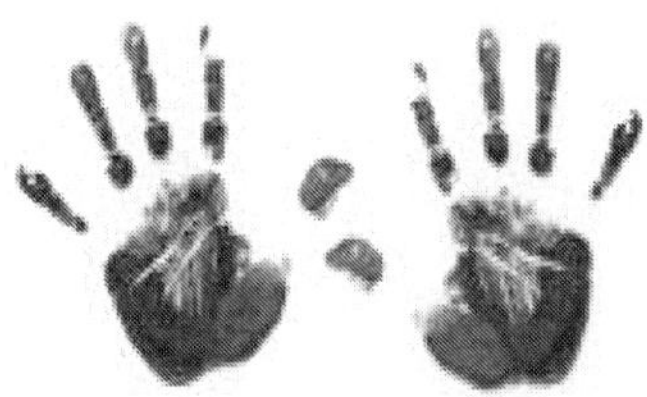

Figure 1-1: Grotter Cosquer, Marseille – circa 27,000 years ago

During the cultural evolution of *Homo sapiens*, the hand modifies and perfects its own operational processes, mediating between elements, adding to existing skills, regulating technical actions, promoting both biological and mental sequential processes (Figure 8).This process follows a pathway, regulating the consistency of technical actions and their relational organisation in space and time. The hand is what humans use to create; the tool determines the space in which Anthropos achieves self-awareness. Although extremely rough, the sketch made by Leroi-Gourhan in *Gesture and Speech* and here reproduced (Figure 1-9*)** is an attempt to illustrate the brain/hand relationship, referring to co-ordinated actions in the manufacturing of objects.

Technology is a creative process connected to human resourcefulness. The emphasis on doing, creating and ingenuity goes back to *teks* theIndo European root of the word 'technology': to weave, meaning to 'fabricate' or 'interweave'. In Greek '*tekton'* refers to a carpenter or a builder and *téchne* to art, ability, skill. Such meanings suggest a process of construction, even creation. During the Middle Ages, the art of weaving, making weapons, ship-building, agriculture or hunting included construction and other productive activities, not simply manufacturing.

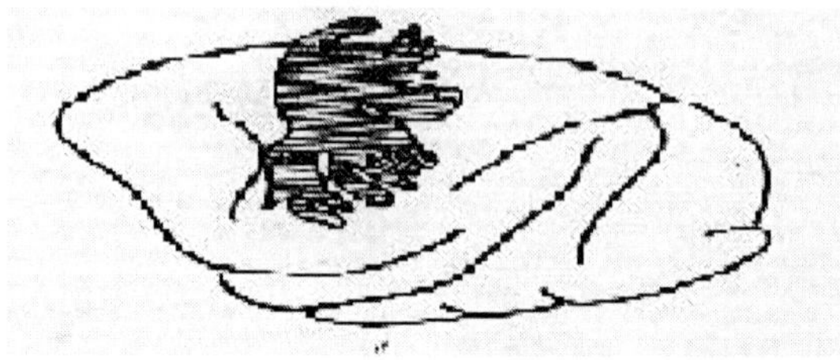

Figure 1-2: The brain/hand relationship according to Leroi-Gourhan (Leroi-Gourhan, 1977, 1, figure 44a)

Technology is the artifice of product consistency, of tangibility, and it determines the uniqueness and quality of interaction. The profound sense of technology is its pragmatic itinerary that allows the body to construct its habitat in the world.[2]

The varied itinerary mentioned above is a study that invites confrontation and participation, not a previously established strategy. It invites the courage to challenge, to reach a consensus between different disciplines and backgrounds. Up until now, these two worlds have been considered as separate - it is now time to take the opportunity to confront each other and perhaps integrate them.

The 'pathway of the tool,' in fact, defines *the order of usage;* the *interface mechanisms* that facilitate the body's orientation in its territorial space, its survival, and the creation of social ties in space and time, functional to the domination of contingent events.

In lithic tools, the hand incorporates the cultural evolution of the first *Homo sapiens* that inhabited the area of the savannah until the emergence of cognitive behaviour, the fundamental characteristics of *Homo sapiens.* Whether the use of nodules of stone to create flints with sharp edges and

---

[2] Defining technology in all its complexity is an arduous task. Technology is usually found in the context of communication, transport, energy or production. The word 'technology' became common usage in the twentieth century after the Second World War. Previously, it usually referred to 'crafts', 'applied science' and 'engineering' to indicate what today is now called technology. According to the *Oxford English Dictionary*, the word was first coined in the seventeenth century when referring to a discourse or a treatise on crafts or industry, cf Bigelow, 1831. In 1959, the *Society for the History of Technology* began publishing a new quarterly magazine entitled *'Technology and Culture'*. For an in-depth analysis of the definition of 'technology' see Dugger, R.L.C: (2002), Itea/Gallup Poll Reveals *What Americans Think about Technology*, International Technology Education Association, Reston, Virginia.

tools for cutting indicates a superior skill or a cognitive capacity with respect to modern anthropomorphic monkeys remains a debatable issue (Toth, 1985, 12: 101-120; Toth et al., 1993: 81-91). The fabrication of lithic tools is the most evident expression of an acquired technical intelligence: having an objective to follow, a project and the final achievement, the tool. It is true that the use of other natural tools such as sticks, stones, and leaves is known to other species, from the otter to the anthropomorphic monkey, but it was the human species that achieved the consequent technological development.

The first humans needed to procure food and defend themselves. In the beginning, therefore, they were food gatherers; alimentation depended on how much nature provided in the way of animals, fruit and roots. It would be reasonable to hypothesise that one of the first problems to solve would have been how to strip the flesh off animal carcasses using implements, obtained from splintered rock and broken bones, long enough to extract the marrow with the help of choppers - partially worked stones.

Reconstruction of the principal stages of the cultural evolution of the genus *Homo* is achieved through analysis and study of rare fossils found randomly and by chance. These finds are very few and far between when one considers the vastness of the area in which they are scattered. Besides, there is the problem of correctly dating and classifying them in chronological order. A more favourable method is to pursue the development of the cultural evolution and the pathways that the various representatives of *Homo* have followed over time, given that in most cases it is possible to analyse the debris derived from their activity. During the middle and later Palaeolithic period, such remains are represented by the leftovers from the slaughter of animals, camp fires and numerous utensils obtained from stone chips, providing precious information based on the type and utilisation of the artefacts collected. A reading of the cultural level can be obtained, perhaps indirectly, and a fairly accurate temporal collocation can be assigned. One of the demonstrated cognitive faculties of hominids is the *intentionality of the tool to be used.* The hominid most likely experienced pleasure in such creativity and this emotion added strength to the lump of stone which would serve to obtain flints from its sharpened edges and stronger implements necessary for cutting the tendons of the animal carcasses. With emotional intelligence the process of techno-social negotiation actively intervenes, integrating sensorial faculties with the site to explore and the pathways to follow.

Pebble culture is believed to be the oldest technology. The first hominids to use pebble chips were the *Homo habilis* who lived in Africa between 2 and 1.5 million years ago. They broke down the stone to obtain a suitable tool for hollowing out, cutting down and making holes. They continued perfecting these techniques until they were able to break off chips from both sides of the stone blocks to have sharp edges for cutting. At first these tools were rudimentary, but gradually they became more symmetrical and regular. The *pebble culture* (sometimes called 'Olduvan' after Olduvai, a site in Tanzania) presents several principal typologies that summarise the technological evolution, illustrated here in figures 3 to 6:

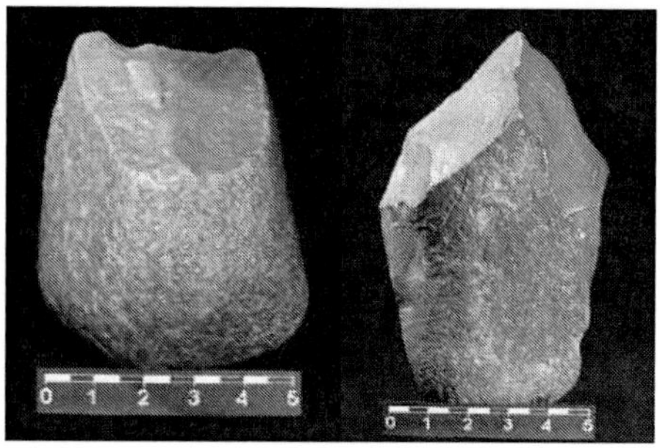

Figure 1-3 (left): Chopperamesrouf, Algerian Sahara
Figure 1-4 (right): Chopping Tool, Tauz, Algeria

(Figure 1-3) Chopper. This is a very primitive tool, a stone chipped away on one side only, to obtain a sharp blade.
(Figure 1-4) Chopping tool. This differs from the chopper in that it was worked on both sides of the stone - 'bifacial'.

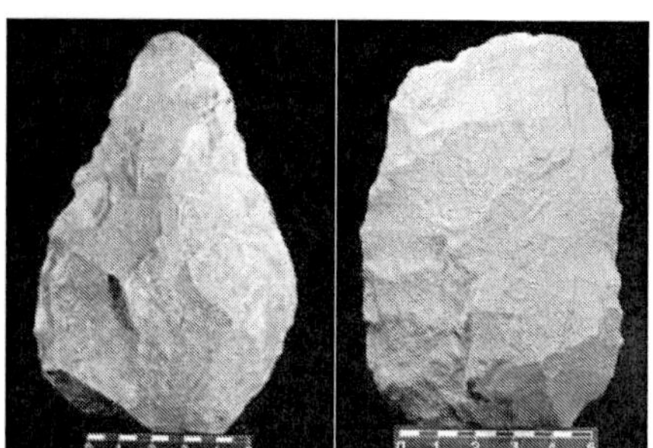

Figure 1-5 (left): *Amigdala*, Gafsa, Tunisia
Figure 1-6 (right): *Hacheraux, Herg Bel Fel Foul, Algerian Sahara*

(Figure 1-5) Amigdala. This is a variety of chopping tool, the bifacial, the 'saved heel'; the stone is chipped away all over. It is called 'amigdala' from the Latin 'mandola' because of its almond shape. The lower left-hand side has been broken off.

(Figure 1-6) Hacheraux.A typical tool of the Acheuleana culture in Africa, the Hacheraux constitute the exception that confirms the rule: they are bifacial, but obtained from a large chip off an even larger block of stone. They normally have a rounded but slightly slanting head.

Today it is known for certain that our direct ancestors (defined by paleoanthropologists as 'anatomically modern') originated in Africa between 200,000 and 165,000 years ago. Genetic knowledge demonstrates that all living peoples today originate from these African ancestors. However, this affirmation concerning the origin of the species of *Homo sapiens* could be destined to be temporary: there is always the possibility that new fossils will be found offering proof that will change current interpretation. For the time being, fossil and genetic proof seem to indicate similar data, meaning that the last model proposed could endure.

According to the estimate of a team of geneticists led by Wolfgang Enard at the Max Plank Institute of Evolutionary Anthropology in Leipzig (Enard et al., 2002:871), it was by means of the process of natural selection that the gene of the human form was established in the population during the last 200,000 years of human history. This occurred following a casual mutation, simultaneously or immediately following the birth of anatomically modern human beings.

The map below proposed by Felman and Cavalli Sforza (Figure 1-7) shows that the migration of the anatomically modern *Homo sapiens* began with a movement towards the rest of Africa around 100,000 years ago and was followed by an expansion, first towards Asia, and then probably south towards Oceania and north towards Europe and America. Along these itineraries the linguistic capacity of *Homo sapiens* evolved in referential manifestations that constitute the basis of modern cognitive behaviour.

In Europe, the indicators of behaviour through language appear unexpectedly around 40,000 years ago when *Homo sapiens* takes the place of the Neanderthals. He invents very sophisticatedstone tools and weapons, establishes a commercial network over long distances, decorates the body with paint and sculpts statues of men, women and animals, shapes animal bones to make musical instruments, and paints images of extraordinary beauty on the walls of the Chauvet, Lascaux and Altamira caves. There is a cultural explosion without precedent in the history of hominids, spreading from Asia to the Cantabrian region of Spain.

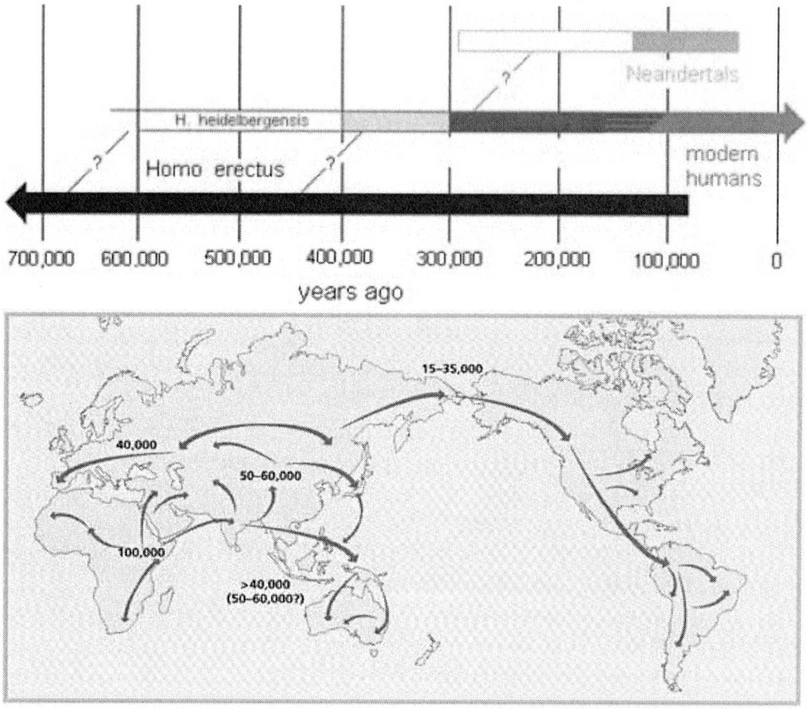

Figure 1-7: The migration of anatomically modern Homo sapiens (Feldman & Cavalli Sforza, 2001)

Archaeological documentation in Africa relative to the crucial period from 200,000 to 50,000 years ago is considerably inferior to Europe where meticulous excavations have been carried out in over a hundred diverse Neanderthal sites. Very few sites have been explored with similar accuracy throughout the whole of Africa. A tie between the African origin of anatomical modernity and the origin of modern behaviour has not, so far, been generally accepted. There is also disagreement about what is meant by 'modern behaviour'. There are two diverse theories: one is that modern behaviour appeared in Africa about 50,000 years ago and is closely related to behavioural modernity revealed in European sites of the later Palaeolithic period; and the other that modern behaviour developed in Africa much earlier than 50,000 years ago and is linked to anatomical modernity through a gradual process which began perhaps 200,000 years ago.

Until recently, the accepted thesis was that the development of modern behaviour was linked to Europe. While genetic and fossil evidence concerning anatomy has been accepted, researchers are still unable to ascertain the contemporaneousness of physical and cognitive behaviour. Several recent African finds seem, however, to predate modern behaviour by thousands of years. In the Congo, for example, denticulate bone harpoons dating back 80,000 years were found; the level of sophistication left nothing to be desired compared to European harpoons of 25, 000 years ago. In Tanzania, small pearls made from ostrich egg shells dating from 70,000 years ago were excavated. Proof of the gradualist theory came to light on the South African coast in the Blombos caves dating from the middle Palaeolithic time. (Henshilwood had discovered a sophisticated bone instrument for fishing there; it was definitely an example of modern behaviour).[3] From an undisturbed deposit of ash and sand, later dated to 70,000 years BCE by luminescence technology, there emerged near the remains of a campfire about ten perforated mollusc shells (Figure 1-8), two pieces of red ochre, five centimetres in length, one delicately engraved with criss-crossing lines to form lozenges, (Figure 1-9).

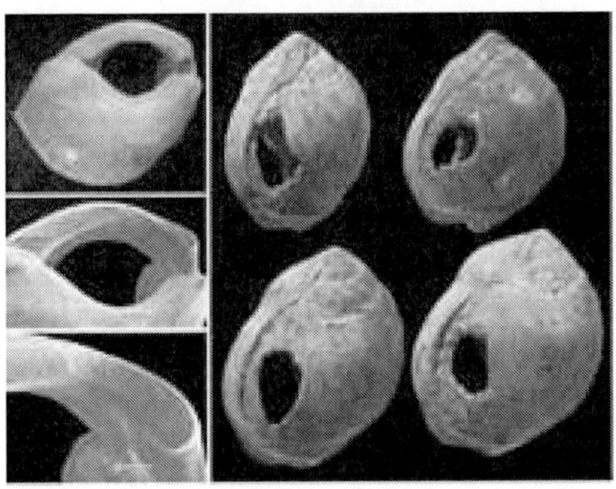

Figure 1-8: Decorated shells

[3] Stone Age, Henshilwood C.S., D'Errico F., Yates, R., Jacobs Z., Tribolo C., Duller G.A.T., Mercier N., Sealy J:C:, Valladas H:, Watts I. Wintle A:G: (2002) 'Emergence of Modern Human Behavior: Middle Engravings from South Africa', Science 15 January 2002: Vol. 295, n. 5558, pp.1278-1280.

Figure 1-9: Blombos Cave, South Africa. Red ochre tablets engraved with lines crossing forming lozenges.

What does the South African discovery suggest? According to Henshilwood, the presence of engraved surfaces denotes cognitive ability and operative capacity attributed to modern behaviour. He hypothesises a patchwork ‘Pan African’ model of development that registers periods of innovation and regression in diverse contexts. In the whole of Africa, there is not one single trajectory towards behavioural modernity; just as there is not one single locality where modern anatomical human behaviour might be documented.

Red ochre, a ferrous mineral, has often been found in Palaeolithic sites dating back at least 100,000 years. It was probably used to protect the skin and to tan animal hides. The tablets in question were first sanded, then ground to create a smooth surface for engraving the delicate crisscross grid, which leads us to consider the presence of an associative convention. This is a fact backed up by evidence: we are in the presence of a tangible product with a relational function that is still to be defined.And yet it may be impossible to establish what this might be, since the tablet is clearly connected to the spatial and temporal context in which it was made. Engraved on reticulated ochre similar to the rock schist, its function is procedural and limited to the context of its use. It could be hypothesised that its function was a point of reference towards the localisation of material space in the reticulated tablet within the system of relationships that bind Blombos man to the context of activity.

1.3 The Explosion of the Species *Homo sapiens*

From the moment *Homo sapiens* makes his appearance in the world, a variety of totally new beings enter stage. His mark of distinction lies in his capacity for technological innovation.This is documented by the sites he inhabited and died in, the caves and graves. The explosion of the species *Homo sapiens* is demonstrated by the wealth and complexity of technological implements, rock paintings and the art of living. The cultural production of anatomically modern man is incorporated into continual exchanges between technique and territory, as documented by the evolution of implements and environmental remains in which location interaction developed, joining physiology to diverse contexts of activity. Within these variable sites, *Homo sapiens'* sensorial equipment built the adaptive processes that made the exceptional biological, and prodigious cultural, success possible.

There are two classes of adaptation. The first implies a pre-programming of the organism capable of coping with the environment. By virtue of such pre-programming, the organism is equipped with all the instincts and skills necessary to operate effectively before receiving any kind of learning preparation. Undoubtedly, this is a most important adaptation mode which enabled an enormous variety of organisms, plants, bacteria and 'simple' vertebrate insects to flourish for millions of years. Such organisms were already equipped with a means of adaptation, needing only a minimum of learning or memory.

On the contrary, the second class of adaptation implied the production of a flexible organism. In this case there was less need for pre-programming and the organism had the task and the capacity to modify its own behaviour in response to the environment. This situation provided for a greater complexity and variety of behaviour. At this point a much larger brain with the basic capacity to learn and remember was required.

The species *Homo sapiens* is the clearest example of a being with these evolutionary characteristics: our capacity to learn and remember enabled us to develop tools, language and technology which in turn increased our skills at communicating and memorising. The capacity to learn and memorise is the most decisive characteristic of our species. Humans share with other animals the biological basis of the brain. Neurons and chemical substances with which these cells exchange messages are the 'bricks of thought' common to many living species. But the same stave cannot be shared to compose the same piece of music; it is insufficient. It is culture that determines what it is that makes us really human. The first anatomically modern humans in Europe are the people of Cro-Magnon.[4]

These first Europeans were hunters and gatherers, living on what they found in their own territory. The Arctic polar cap extended as far south as northern Germany and southern England. Below the ice cap there were vast plains void of vegetation, but frequented by large herbivorous mammals. For skilled hunters gifted with the cognitive capacity of *Homo sapiens*, the abundance of fauna that populated the vast steppe was a great nutritional resource for the taking, often requiring very little effort. According to archaeological documentation, they camped near rivers and, in particular, in places near to where herds of mammals such as reindeer would attempt to cross to the other side, vulnerable to attacks from hunters. Rock paintings show that these new inhabitants were knowledgeable about how animal fur varied according to the seasons. For example, the bison moulted at the beginning of summer, deer in autumn and the rhinoceros showed a pleated skin visible only in summer. In fact, it is thanks to the rock paintings of the upper Palaeolithic period that the anatomy of many of the animals is known, since a number of them are today extinct.

In the currently available paleontological archives, covering the period from 40,000 to 10,000 years ago, there are indications of innovations which have never been published. The camp sites of the Cro-Magnon are quite different in size and complexity from those of the preceding periods: overhanging rock shelters generally faced the morning sun and dwellings built in open-air camp sites were complicated structures. Those worthy of note were built 17,000 years ago. In the Ukrainian site, Mezhirich, the remains of four dwellings in the midst of a profusion of neatly lined-up mammoth bones were found. The ordered and varied way in which the bones had been selected and put away led to the conclusion that such constructions were the 'first examples of architecture' (Tattersal, 1999: 280). Fire, already mastered in a rudimentary fashion by the Neanderthal, is used by the Cro-Magnons in more varied ways. They built lamps to light up the inside of caves, and their fires in different styles were complex; it seems that they used heated stones to boil water in holes lined with animal skins.

---

[4] As is well known, the man of Cro-Magnon is an ancient variety (cromagnonoid) of an anatomically modern human being who populated Europe extensively during the late Ice Age. The oldest remains found by French archaeologist Louis Lartet are dated around 30,000 years ago just before the Combe-Copelle discovery; it is considered a variation.

Around 28,000 years ago, in a site found in the Czech Republic, Cro-Magnons baked small clay statues in ovens heated up to 400°. Their stone tools reflected their unequalled technological skill. The basic technique consisted of shaping a large stone block into a cylindrical form so that with the possible aid of a wooden hammer or bone, long thin blades could be broken off. Such blades had long sharp edges and could be modified using more specialised tools. In this way, it was possible to obtain a useful blade length, using ten times more per kilo of natural material than previously used techniques. The points of antlers were chiselled off to make them straight for spear throwing; often they were decorated or engraved with intricate and refined patterns. Used also by Eskimos and Australian aborigines, the spear end was placed in the thicker end of a baton which served as a launching base. The huntsman gripped it at the narrower 'head' to aim at his prey, thus increasing thestrength of the arm, enablinghim to throw the spear further and with greater precision. The Cro-Magnon had already perfected and diversified his fishing skills 20,000 years previously with serrated harpoons, often with a canal running down the centre to drain off the blood of the prey. Clay moulds dating back to the same period indicate how plant fibres were used to make ropes.

From the oldest cultural *facies* of Cro-Magnon man and the Aurignacians, human groups settled in territory using technical implements. The complexity of dwellings, open-air camp-sites without natural shelter located near running water; undercover shelters – caves and hanging rocks – furnish the most important proof regarding the most recent stages of cultural development. Furthermore, the long-lasting sequences of the occupational phases of these sites have served to set the framework on which most contemporary research is based. In no matter what type of climate, but especially cold, natural shelters were much sought after, a fact which has proved invaluable for research. What worked well for one group of inhabitants also worked well for another group over thousands and thousands of years, each leaving a layer of debris for those who followed in their path. These strata of debris reveal the cultural organisation of successive inhabitants, making it possible to order, with precision, related elements of the various settlements. The technical systems revealed in upper Palaeolithic times constitute a local *combination platform* between preceding and successive innovations. Each new invention (a tool for work, a musical instrument, a wall painting)is a creation generated by a selection of the preceding heritage of memory, new variables characterising the context in progress.

Such uses might be regarded as a Palaeolithic thread, regulating the motility sense of the body through contingent contexts dependent on the most elementary means necessary to all humans – prehension and the many types of percussion. It is the hand that is the main element, involving the whole body. Probably it will never be possible to establish the precise moment in which humans began to make implements. Each single tool may be made in many different ways, and, often, such different techniques indicate distinct traditions in industries characterised by similarities. The treatment of tool-making afferent to such necessities privileged utensils in stone, which in many cases are all that remain of the activities of a particular group.

Right from the beginning, humans were, however, capable of using a great variety of primary materials; only rarely did they preserve organic material. The bone, as well as the closely related horn, is a material that has been well preserved, especially in caves. In several periods in European pre-history in which it was very cold and wood was probably scarce, bone and horn had multiple uses and were chipped or hammered into the shape desired. Alongside purely functional objects, especially in later periods, there are abundant examples of personal ornaments: engraved bone pearls, seashells, necklaces of perforated animal teeth, earrings of various materials. Buried skeletons have been found with pearls attached to craniums and scattered among thoracic bones. These pearls were probably sewn to a cap and on a tunic similar to the leather clothing worn by Native Americans.

Since very ancient times, colouring substances derived from natural earth pigments were utilised for cave painting, but it is possible that they were also adopted for cosmetics and body paint as used by Australian aborigines today.

In the oldest Aurignacian sites dating back 30,000 years, musical instruments have been found: bone flutes that can produce a remarkable range of notes, and identifiable percussion instruments. Among the many carved objects brought to light is a stone tablet dating back 34,000 years from the Périgord region in France, identified as a lunar calendar.

Regular graves for the dead were discovered for the first time at Cro-Magnon sites. The most surprising example was unearthed in Russia at Soungui. 34,000 years ago, a man was interred with two young females, wearing material signs of great wealth. Over three thousand engraved ivory pearls were found, perhaps sewn to their clothes. Each wore a necklace, earrings and bracelets made of shells. They were buried, their heads touching; by their side were weapons measuring approximately two metres in length. This refined grave is one of the most spectacular examples of the Cro-Magnon civilisation. The level of complexity of each diverse burial reveals considerable variation denoting both social stratification and a division of labour within this society.

The Magdalenian culture began circa 20,000 years ago. During this period, the most relevant wall paintings were executed in the caves of Lascaux, Altamira, Les Trois Frères, Font de Gaume, Les Comberelles, Chauvet, Cosquer, Cusac, Rouffignac and Niaux.[5]

The fact that paintings have always been found inside caves is a most fortuitous circumstance which has favoured their conservation down to the present day. However, these works are probably only an infinitesimal portion of the entire artistic production lasting over thousands of years. Three colours were used: yellow, red and black, but it was possible to obtain a great variety of combinations as is evident in Font de Gaume and in Altamira. Yellow was obtained from egg yolks, red from iron oxide minerals and black from magnesium dioxide. These were applied to the walls with the finger tips to draw a line; the whole hand spread the colour to cover a larger surface. To achieve the stencil effect (Figure 1-10), and the Pech Merle horse prints, the colour was blown through a short bone tube covering the hand to obtain a 'negative' image (Collins, 1980).

Moving further north east towards the Urals and Scandinavia (Figure 1-11), rock paintings were found in deep caves. Probably, this type of painting was applied to rocky external surfaces similar to the more recent African painting. Over the course of twenty thousand years or so (the period of pre-historic art in Western Europe) it is possible to observe the development of diverse techniques which, however, do not correspond toa region and have no chronological order.

Cro-Magnons lived in the entrances of caves, where there was shelter and light. *Homo sapiens* lived a culture of light, day and night. The characteristic uniting Cro-Magnon cultural practice is the *stratification of techniques* to produce carvings and paintings. Wherever a site was inhabited, either permanently or occasionally over a long period of time, decorative stratification is evident. The more recent works covered the more ancient, providing a series in sequence, clearly showing the cultural changes which had occurred over the course of time. For example, in the dwelling places, objects such as small statues are associated with stone implements, a fact that facilitated the establishment of the Palaeolithic period according to *changes in techniques*: figurative representations in

[5] The existence of rock painting in upper Palaeolithic times is by now an accepted fact. There is direct evidence of many remains under rocks, and in caves spread throughout Western Europe with which we are familiar - the painted figures, the carvings and the sculptures. It is situated 200 metres from the original caves displaying replicas of the great hall of the bulls and the painted gallery.

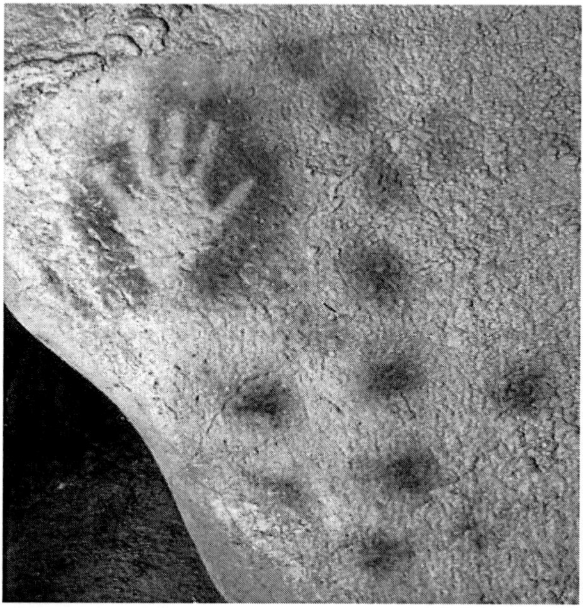

Figure 1-10: Hand 'negative' created by blowing paint through a bone tube, Pech Merle, France.

Figure 1-11: Wall painting in the Stora Backa caves, Sweden

the caves, the accumulation and conservation of objects left over from a former material life, and the skeletal remains of humans and animals. During upper Palaeolithic times, innovative technology in Europe developed in successive spurts: bones appear around 36,000 years ago, bone needles around 28,000 years ago, serrated harpoons around 20,000 years ago. The practice of kilning clay pots at high temperatures began circa 30,000 and 24,000 years ago. There is an explanatory sample found in Dolni Vestonice, in the Czech Republic where female statuettes were baked and afterwards smashed to pieces. The baking of clay pots was practiced intermittently over a period of 150 centuries, until Neolithic times. It marks the beginning of ceramic manufacturing for utilitarian purposes.

It is possible to speak briefly of the 'emergence' of *Homo sapiens* when paleo-ethnological research is able to document the variable combination between ancient and modern in the evolution of techniques marking a major discontinuity between humans and other animal species at a phylogenetic level. But how should Cro-Magnon man be considered?

Above all, the facts show that when *Homo sapiens* arrived in Europe he already had the cognitive capacity of today. According to the English archaeologist Steven Mithen (2007:309), it is 'cognitive fluidity' generated by language that separates the mind of *Homo sapiens* from his predecessors. He had the capacity to integrate thinking processes and accumulated knowledge, derived from distinctive intelligence, to create forms of thought that would never have been able to exist in a domain-specific mind. In *The Singing Neanderthals: the origins of music, language, mind and body* (2007), Mithen proposes pre-linguistic communication as the evolutionary basis of language. Referring exclusively to hominids in *The Prehistory of theMind* (1996:270), he affirms three basic intelligences emerge from the documentary production of hominids: social intelligence to manage their habitat and their surroundings; intelligence regarding their understanding of the historical and the natural world of animals, plant life, climate and seasons; and technical intelligence to manufacture tools and objects in stone.

On the basis of this observation, Milthen argued that the 'cognitive fluidity' of language overtook the 'multiple' intelligence and 'domain specific' and constituted the 'mark' of *Homo sapiens*.He supported the research of American philosopher Pierre Carruthers, emeritus professor of philosophy at the University of Maryland, and author of *The Cognitive Functions of Language*, published in the journal 'Behavioural and Brain Sciences' (2002, 25:657-726). Milthern recognised the evolutionary bifurcation that led from the musical protolanguage of hominids to the system of cross-modular integration that characterised the interface between languageand cognition. Hominid protolanguage ('Hmmm') was a system of 'holistic' communication in the sense that the composition of multisyllabic expressions could not be broken down into meaningful subgroups; 'manipulative' in the sense that it stimulated into action or provoked particular emotive states; 'multimodal' in the sense that it involved diverse cognitive and emotive models; 'mimetic' since it imitated natural sounds and animal cries through relations of synaesthesia.

It would seem that Milthen overlooks the proven fact that evolution mainly concerns adaptive strategies and differential reproduction, until offspring are in a position to better survive in successive generations. Long-term survival offers the possibility of constructing wider social networks and a heritage of more extensive knowledge. Intelligent brains have a lot to do with better survival strategies. This does not mean denying the importance of language, but rather a particular type of action in the world: an attempt to elaborate and modulate interaction and coordination between humans. Mithen's thesis is undoubtedly an excellent one for its remarkable documentation, but it is prisoner to the strong philosophical tradition that considers brains as if they were sophisticated computers, elaborators of information, cognitive machines. This is clearly a methodological ingenuousness before complex notions such as 'gene', 'body', 'emotion', 'brain'. It is convincing to argue that the passage of *Homo sapiens* was marked by both a growth in motility heritage and the more articulated capacity to intentionally communicate through hand gesture, often accompanied by vocal expressions. Neurophysiologic data (Liberman and Whalen, 2000; Watkins et al., 2003) indicate that the long evolutionary journey towards language was marked by a series of decisive stages (integration of an oral-facial system with a manual system, the formation of an arsenal of pantomimic proto-gestures, the appearance of a bimodal protolanguage of gesture and sound, and finally, the appearance of a prevalently vocal system), each one of which seems connected in turn to a developmental phase of a communicative mechanism based on the ability to act.

Such a 'special' capacity facilitated flexibility and adaptability to diverse environments based on equipment from the territory obtained over time. Levi-Strauss (1970:30) defines it as 'a set of instruments and material collected over time'. What is more anomalous, given that the composition of this set bears no relationship to the project of the moment, or for that matter, with any project in particular, it is the contingent result of whatever occasion has arisen on which the stock might be renewed or enriched or maintained with building from leftover materials or prior demolitions.

1.4 The Bond between Man and Time at the Dawn of Working Memory

The beginning of the upper Palaeolithic is marked in Europe by the substitution of the Neanderthal by anatomically modern humans who probably originated in Africa. For some time it has been a widely held theory that the extinction of the Neanderthal by modern human beings was due to their intellectual superiority. However, there is a growing amount of data which indicates that the Neanderthal practiced many complex activities that once were considered characteristic of anatomically modern humans.

Characteristics	Common	Occasional	Absent	Uncertain
Art				✓
Use of pigments	✓			
Jewellery		✓		
Symbolic burial of the dead				✓
Exchanges over long distances				✓
Microliths		✓		
Serrated points			✓	
Bone implements		✓		
Blades		✓		
Needles			✓	
Exploitation of marine resources		✓		
Bird hunting		✓		
Division of labour			✓	

Table 1-1: Proof of modern behaviour among the Neanderthal (Finlayson and Carriòn, 2007)

Neanderthal remains reveal that at least several groups were familiar with tools and knew what they needed to know about the wide range of food available. Nevertheless, these practices seem to have been more widespread and more rooted in modern humans than in the Neanderthal, perhaps giving *Homo sapiens* an advantage over them.

The advent of modern genus *Homo* is a biological revolution, but it is also the manifestation of a number of technological and behavioural innovations. In the list of utensils, the development of technology was transformed by the use of hard materials from dead animals (bone, ivory and antler). The development of differentiated industries and the rapid rise in the level of standardisation of utensils and weapons marks the beginning of the upper Palaeolithic period, while the emergence of *operative chains* made up of more technical movements is considered an indicator of complex cognitive processes. In Europe, from about 40,000 years ago, changes began to occur in both the technical and social spheres, but the upper Palaeolithic 'cultural revolution' was most probably a linguistic one as well. The transport of material from natural deposits to sites where it would be used represented quite a different type of social organisation from previous social systems (Feblot-Augustins, 1993). During this period, well defined ethno-cultural groups came and went in fairly rapid succession through certain territories and were organised into *active networks over great distances*. 'Industries' revealed by Palaeolithic research are the recurring outcome of an active interlocking process between human body movement and action. Bateson (1984:123) referred to this process as 'levels of aggression'. While interactive fields were generated by the continuous synchronisation of human movement, such levels determined the trajectories of use and the selection of material (wood, stone and bone) according to the continuous perceptive monitoring of the task at hand.

The experience of Palaeolithic man depends on context, needs and adaptation to the dynamics of the environment with the implements at hand: hunting for food, searching for a better environment, making clothes and building shelters. The complexity of living structures, located in the open-air with no natural shelter, usually situated in proximity to water or in caves or under natural overhanging rock, provided the most important evidence for recent studies on cultural development as well as the long sequences of occupational phases in these sites setting the framework on which to base the greater part of contemporary research. In no matter what type of climate (mainly cold) natural shelters were highly sought after, a valuable factor for students in the field. What was suitable for one group of inhabitants was also suitable for other groups over a period of thousands of years, each one leaving layers of debris on top of what had been left by the preceding inhabitants who had in turn camped on what had been left by those before them. These layers of debris reveal the cultural scheme of successive occupants and further permit an exact study of elements relative to the various occupations. The cultural sequences defined inside the caves also serve to date isolated material coming from open-air settlement sites.

Techniques revealed in upper Palaeolithic settlements denote a local combination platform between preceding innovations and successive techniques. Each new invention, whether a work tool, a musical instrument, or a wall painting, is a creation generated from a selection of preceding creative wealth where new variables characterise the context at hand. For the Palaeolithic human it was a period of heterogeneity and the unexpected, modulated by the selective attention that governs adaptive processes. On a par with other animals, Homo sapiens inhabits his own sensorial world and exercises his own activities through contingent territorial events. What makes him different is the invention of tools, that distinct expression of evolutionary genes. Within this picture in progress, technique functions as the measure of transaction between the rhythm of living material including the human body, and the specific material variables *hic et nunc,* the context. The tool is in fact the synergetic bridge between the biological sequence and the adaptive synchronism involved in the technical execution.

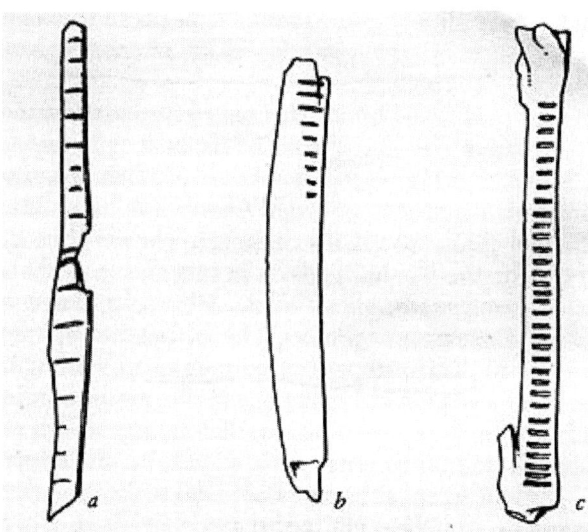

Figure 1-12: Series of parallel incisions on a bone or stone dating back 35,000 years, Chatelperron, France

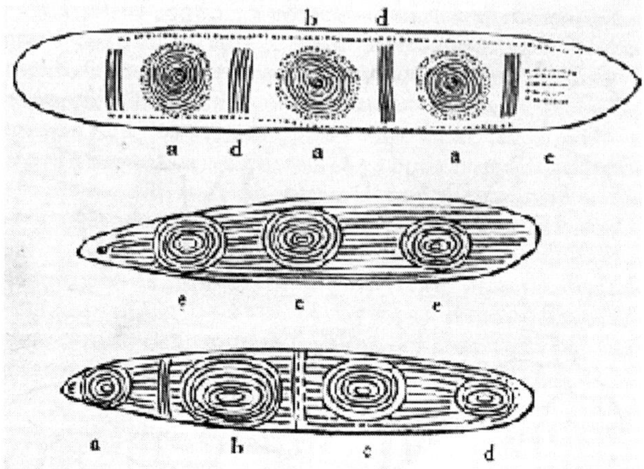

Figure 1-13:Australian Tjurunggas

Already by the end of the Palaeolithic, tight rhythmic ownership signs with tangible limits are in use between men and their tools, for example the return of a particular wild fowl or the ripening a particular fruit. Earliest evidence is found in the rows of cupules or incisions on bones or stones (Figure 1-19) which Leroi-Gourhan considers 'proof of the oldest rhythmic manifestations ever expressed (1977, I: 222). The French Palaeontologist likens such incisions at regular intervals to Australian *tjurungga* (or *tjuringga* or *churinga*). They appeared towards the end of the Musterian period and increased in number by the end of the Chatelperronian period (Figure 1-20). The *tjurungga* are ellipsoidal blades in wood but are more frequently found in stone, are at least fifty centimetres in length, and engraved or painted with circles, spirals, nets or groups of dots (Strelow, 1907, 20, I-V). Like the Palaeolithic incisions, the *tjurungga* decorations carried out by the Aranda and Lorija cultures from Australia are organised along a diachronic axis, and according to Obermeier (1916:21) converge in a recitative function.

The incisions in question demonstrate the evolutionary emergence of the manipulation of utensils; transactions of single temporal duration *in progress*, responding to a specific activity, (for instance, hunting). Their 'practical use' focused on mediating relations between the human agent and time. Each single incision is different from the next along the diachronic axis representing a long period of duration. They constitute a most effective sample of the bond between *man and time*, distinguishing *Homo sapiens* from any other species.

Considering the above incisions referring to the notion of time, Leroi-Gourhan (1977, II: 364) defines the latter as 'a human reality par excellence.' This definition focuses on 'the two measures of time', in which methods determine the theme of classical science. The first refers to the measure of time, chronology, localising events along a graded scale so that events might be classified according to the order in which they occurred.

The second refers to the duration, the time interval between a sign indicating the beginning, and another indicating the end. Adhering to the 'world-clock' mechanistic model,[6] Leroi-Gourhan (1977, II: 371-372) considered time 'parallel to biological evolution'.

Time is seen to be synchronically on a parallel route, turning into a grid behind which individuals are locked when the system of relationships reduces the terms of transmission to hours, minutes and finally seconds.

Current scientific prospects offer a very different scenario. Irreversible processes have put back into play notions of structure, function and history. In this new perspective, irreversibility is a source of order and organisation. Prigogine (*Reference ?*) observes that 'the role of science is to find links; time is one of these links. Man is a product of time; if, on the other hand, man created time, time would evidently be a shield between man and nature'.

There is a biological history of time. Time is the irreversible evolution of the species. Time came into existence with life millions and millions of years ago, passing from generation to generation, from species to species, but that is not the whole story. It became more and more complex along the way. Homo sapiens did not appear on earth until 'one second ago', in a geological sense, that is. Evolutionary theory foresees this result on the basis of the principle that the nervous system progressively increases in complexity. The 'emergence' of human beings was rather the fortuitous consequence and contingency of thousands of connected events any one of which could have occurred in a diverse manner deviating the course of history to an alternate route that would not have led to a human type of intelligence. But, to understand events and the generalities of the course of life, it is necessary to go beyond the principles of evolutionary theory. It is necessary to analyse the Paleontological documentation of *the progressive contingencies of the history of life* on our planet, the only version that has

---

[6] The thermodynamics of complex systems reversed the 'world-clock' mechanistic model showing irreversibility to be the font of order and organisational creation. The organisation of living beings and the history of man are not casual accidents of the universe, as Jacques Monod argues in *Chance and Necessity: An Essay on the natural Philosophy of Modern Biology, New York, Alfred A. Kopf, 1971.*

actually came into effect out of the thousands of possible alternatives. A similar concept of the history of life strongly contrasts with deterministic conventional models of western science, i.e that the human being is considered to be the highest expression of life, whose destiny it is to oversee the planet.

The 'emergence' of *Homo sapiens* from the 'irreversible time of evolution' constitutes a source of biological organisation. The formation of culture has certainly not posed a limit to the biological evolution of humanity. On the contrary, it constitutes an element of support for its continuation. Cognitive and social behaviour that increase the cumulative value of the human population tend to be reinforced by epigenetic limitations, which are in turn controlled by genes. Both biological and cultural evolution are characterised by similar possibilities: the former to transmit genes (with modifications); the latter to transmit (with modifications) the functioning of acquired skills through trial and error. Transmission with modifications is the essence of the capacity to evolve. The capacity to transmit inventions that have been produced and proven has made it possible to establish a society capable of passing on traditions and innovations.Those transmitted traditions and innovations have facilitated the multiplication of capable individuals as opposed to other contemporary human groups. Perhaps this is the way a new species emerges and is successful: it settles and accumulates a series of rich mutations full of evolutionary potential over a long period of time. They supply the organism with the necessary original characteristics that make adaption to the environment of everyday life possible. Those characteristics improve capacity to survive and reproduce, which in turn causes a demographic explosion usually beginning in a small region and then continuing to expand for thousands of years.

The process of expansion of anatomically modern man was very slow to begin with, but today we are beginning to understand the significance of time and place. Man was born a *technologist*. In the beginning he created rough tools; they were so elementary that it was not clear whether seemingly antique bowls were utensils made by man or the result of some accident of nature. Slowly these implements were modified and improved and modified again until they could no longer be considered accidental products. Such a development incorporated technological inventions and constituted the engine behind the success of anatomically modern man. Cave wall paintings, bone instruments, body decoration, burial rituals, increased hunting activity, exchange of goods over long distances and structures in camp sites all document *topographical transactions* inwhich technique is incorporated with implements belonging to a temporal order of 40,000 years ago, when *Homo sapiens* begins to take over and replace the Neanderthal.

Executive functions controlled by the pre-frontal cortex in technical activities are already developed. According to Goldberg (2004:94) 'the selection of necessary information for solving contingent problems is carried out by the pre-frontal cortex which needs to know, at least more or less, where this information is stored in the brain'.

The final conquest in the evolution of the nervous system is the pre-frontal cortex and it is only human beings (and to a certain degree anthropomorphic monkeys) who reach such a level of development. The pre-frontal cortex sets us apart as social beings in so much as it regulates the systematic fabrication of implements and the functional ability to adapt and learn. The fundamental capacity of the pre-frontal cortex is to find 'executive' solutions for totally new situations. The complex nature of executive functions includes multiple capacities such as concentration, flexibility, the ability to cope with the new, and above all 'adaptive resonance'[7], the capacity to recognise a repertory of experiences, of memory useful for coping with novelty. Unlike other species, individual humans do not have to continually rediscover the world of their forbears right from the beginning every time. They benefit from the incremental effect of knowledge that society has gradually accumulated over thousands of years. They have stored and communicated knowledge in symbolic form through various cultural systems, transmitting the known from generation to generation. If knowledge is defined as the availability of a rich repertoire of models that facilitate the recognition of new situations and new problems as if they were familiar, then human beings are truly a unique species. The capacity to transform the world through the heritage of memory interacts with the executive function of the pre-frontal cortex to create direct reactions to meet the new. It is the technical implement that incorporates this capacity and distinguishes *Homo sapiens* from all other species, in as much as he is a communicating species defined by the ability to use technology. The technical instrument is, in fact, the cognitive anchor of the human species; it models man's bond with time, deeply rooting the sensorial body in the physiology of the instrument. At the dawn of working memory it is the hand of *Homo sapiens* that leaves its mark on places modelled by technological innovation.

---

[7] This type of recognition involves enormously different temporal scales ranging from millions of years to only a few. Different species or groups of species depend on different repertoires of phyletic memory for their survival (Fuster, 2003).

CHAPTER TWO

THE RHYTHMIC BODY: THE LOOM OF WORKING MEMORY

2.1 The Mask of Greek Tragedy

Research on memory and developments in the neurosciences have demonstrated that human beings automatically and unconsciously organise their personal experiences into spatial units. These experiences determine the limitations established by mutations in the territorial location of the body and its relationship to time and interactive elements. The procedural capacities that support the evolutionary history of animal intelligence set the mnemonic system of human beings in motion. This mnemonic system is generally manifested in the form of habit, perceptive strategies or motility, and associative or non-associative types of conditioning. Such complex and variable processes constitute implicit (or procedural) memory, a form of memory that is not recalled at a conscious level[1]

Conceptualised by French philosophers in the eighteenth century (Maine de Biran, Ravaisson and Bergson), memory does not only govern simple perceptive and motor capacities such as the leg movement of a

[1] According to Squire and Shachter's (1987) definition, what we usually consider conscious memory is today called *explicit (or declared) memory*. It is the conscious memory of people, places, objects, facts and events. Conversely, we term implicit (or procedural) memory, unconscious memory that concerns addiction, sensitisation or classical conditioning, and perceptive and motor ability such as riding a bicycle or serving a tennis ball. Philosophers and psychologists had already anticipated this distinction between explicit and implicit memory. Herman Helmholtz, the first to measure the conduction of action potential, also worked on visual perception. In 1885, he established that the greater part of mental elaboration for visual perception and action occurs at an unconscious level. In 1890, William James, in his classic *Principals of Psychology*, extended this idea to addiction, (unconscious, mechanical, reflexive action) and to memory (conscious awareness of the past). Not only as conscious memories but also as unconscious memories. The latter are usually inaccessible to the conscious mind, but sometimes they, too, exercise a powerful effect on behavior.

cyclist, Margot Fonteyn's pirouettes or the techniques of Wynton Marsalis' trombone. Descartes observed that when playing the lute, the musician had memory in his hands. Implicit memory constitutes the biological basis of any performance, guiding us through well-rehearsed routines - the abilities and habits that have already been acquired with experience. If explicit knowledge can be communicated through language with a high degree of awareness, implicit knowledge fundamentally concerns behaviour, emotion and soma, referring to the 'here and now' of the present moment (Stern, 2004).

Memory is transformed and established according to its content, and as Freud intuited, the dynamics of these transformations are influenced by an individual's experience and the specific environment lived in at the time. Through trial and error, a technological selection is made, creating a useful repertory of memory, oriented towards the environment. De La Jara's (1990: 199-217) studies demonstrated, for example, that the Incas employed a complex writing system using elements such as cloth and painted vases that served to record chronologies, calendars and chronicles. This relationship between different materials (cloth and ceramic) was defined by Cardona (1981: 36) as 'multimedia', that is, various materials were used and were adapted according to needs. One system, known as *quipu* (fig. 1), recalls the usage of knotting a handkerchief as a reminder of something; such coherent cultural structures pre-dated the invention of writing[2]. Up until recently, however, the *quipu* has been regarded as a simple accounting method or a mnemonic system.

Referring to the subjective experience of loss and death, memory was not always the *maison manquante* (Boltanski, 1995) of oblivion; neither was the magic looking glass an object of superstition and trickery, in contrast with what Francis Bacon (*Novum Organum Scientiarum, 1620*) recognised as the thoughtlessness of human rationality. Similarly, the 'non luogo' where one idly absorbed 'the simulacrum of the world' confused by

[2] What is surprising when consulting the numerous studies about writing in the West is the idea that the various systems are phylogenetically organised towards evolutionary perfection. The alphabet is the final phase. Studies regarding the evolution of writing are in fact histories of the alphabet on what has gone before; the attitude still prevails today in the classification of languages and social institutions. The alphabet is treated as a technical invention, classifiable in terms of 'progress.' From an anthropological and semiotic perspective – a field in which writing and the graphic use of it are for communicative ends - it makes no sense to speak about evolved forms. Every society uses a type of writing that it finds congenial to its needs, or it adopts it because of external pressures or cultural forces, integrating it with other symbolic systems.

'the crystal ball and the smoke' (Changeux, 2000: 93) was not always without structure either.

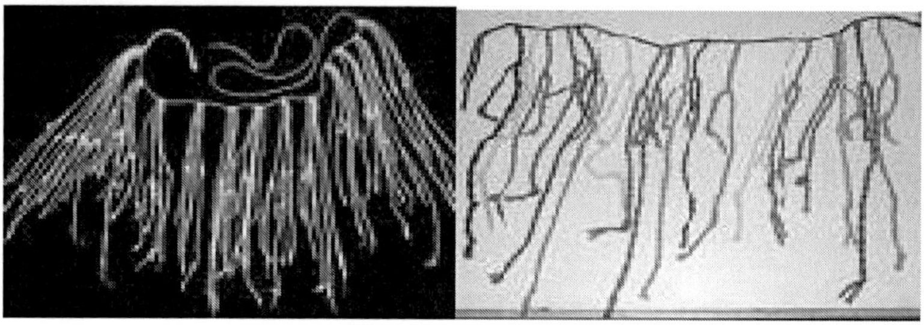

Figure 2-1: Two Peruvian quipu held in the Museum De la Nacion in Lima, Peru

Frances A. Yates (1972: 5-6) formulated a fascinating hypothesis in her book on the art of memory. In the pre-modern world, almost totally without instruments to facilitate memory, there probably existed, though lost long ago, a 'creative technique', a *memorisation faculty*. Experimental anthropologists and psychologists Lurija (1952), Bartlett (1968), Watt (1968), Vansina (1978) and Ong (1982) demonstrated Yates's hypothesis in studies carried out on pre-modern societies with oral traditions. For example, when Bartlett (1932) asked individuals involved in an oral mnemotechnic experiment to reproduce a complex story or an argument, he noted how precision was very rare in the account. Such individuals *reconstructed* mnemonic material instead of *reproducing* it. The process became particularly evident when they were asked to give an account after a lapse of time.

From this study some surprising observations emerged. An *active organisation* of the experience was generated rather than a reproduction of a word for word account. Lurija also carried out extensive research on illiterates and subjects of low literacy levels, in remote areas of Uzbekistan and Kirghizia in the Soviet Union of 1931-32. He demonstrated that oral thought is constructed along practical itineraries connecting situations to human senses. The illiterate subjects under examination identified the geometrical shapes administered to them with concrete objects they had experienced. A circle, for example, was called a moon, a sieve or a basket; a square, a door, a house or a board for drying apricots. Similarly, drawings of a hammer, saw and axe were exclusively associated with a block of wood representing functions of daily life. From such experimental analyses, memory emerges as an activity in which the technique is modelled according to the material, it measures ways that constitute the adaptation of man to his own ecological niche.[3]

Oral exercise is above all invention *in situ*. In oral-based societies, memory is established through circumstance, responding to what happens – a surprise becomes an event. Michel de Certeau (2005:136) observed that:

> Memory deteriorates when it can no longer react. On the contrary, thanks to independent events, memory is based on the expectation that something unrelated to the present will happen or should happen. Far from being a shrine or trash bin of the past, memory believes in the possible, it is full of expectation, keeping watch - lying in wait.

Memory is a journey of human emotions, a technological journey. Modelled on experience, it constitutes the basis of human identity. We live in the world through memory, which deposits the culture of belonging through our body, the experiences accumulated in the family we were born into, the emotional expressions acquired. Together, along a path hampered by life's difficulties, they enable each one of us to create our own biography, the uniqueness which unmistakably makes us who we are. The implicit actor in this 'life' is the 'body in action' in the environment. Its function is recognisable in the weaver's shuttle as it moves to and fro, across the loom between the internal surface (the relationship with the whole body) and the things that make up the relational frontier between the individual and the environment – bridges that connect individual stories with the community.

The art of memory has an ancient tradition. Aristotle spoke of those who control 'the imagination mediating between the will and fabricated images that fill the places of mnemonics'. Cicero, Quintilian, Albertus Magnus and Thomas Aquinas all made particular reference to this art, theorising the rules in great detail. These rules have been expounded with more sophistication by a number of modern thinkers. Up to the time of Leibniz, an educated person was aware of these theories, but over time, interests in the argument varied. In the beginning they were conceived as *neutral techniques* to be used in rhetoric and *ars praedicandi. H*owever, they soon became loaded with metaphysical meaning. After the traditions of the Spanish philosopher Raymond Lull, the cabalists and hermetic

[3] For further reading see:
- Malinowski, Bronislaw Kasper, Scientific Theory of Culture and Other Essays, University of Carolina Press, 1944;
-Radcliffe-Brown, Alfred, Structure and Function in Primitive Society: Essays and Addresses, 1952.

thought, the art of mnemonics was totally transformed into an encyclopaedic undertaking, a total 'world theatre'.[4]

The subjective experience of oblivion belongs to modern thought. Without cultural flexibility and social function, memory moves into literary territory, the art of 'narrating time' mirroring and metaphorically symbolising Cassirer's (1968: 80-81) definition of man – an *animal symbolicum* who cannot avoid the conditions of existence he himself created.

Man no longer stands before reality; he can no longer see it face on. Physical reality seems to recede the closer the symbolic activity of man advances. Instead of relating to real things, man in a certain sense is always in dialogue with himself. He has surrounded himself with linguistic forms, artistic images, mythical symbols and religious rights to such an extent that he would no longer able to see or know anything if it were not for this artificial mediation. The same situation is to be found in theoretical and practical fields. Even in the practical field, man does not live in a world of simple facts according to his needs and his immediate desires. Rather, he lives between emotions aroused by his imagination, between fantasy and dreams. According to Epictetus, what disturbs and upsets man is not caused by things, but rather by man's opinions and imagination about these things.

In the modern sense, the 'subject' is essentially an observer, but he becomes 'the 'object' of both his environment and himself in an analytical capacity (Taylor, 1989: 143, 176). Observation implies detachment and ability for abstraction. The result is cognitive confidence and rational control. At the beginning of modernity, the birth of subjectivity relies on concepts such as *mind, consciousness, hegemony of reason or radical reflexivity* avoiding oblivion (Taylor, 1989: 153), through memory. The Cartesian subject exists in the measure in which humans think and achieve

---

[4] In the philosophical traditions of the West two radically diverse notions are present - oblivion/forgetfulness and memory/recollection. Both go back to ancient sages and the related literature is endless. To simplify the debate, the first notion perceives forgetfulness and oblivion as definitive loss, the tentative ideas, images, notions, emotions and sentiments present at one stage in the individual and collective consciousness become absent. This notion became progressively more complicated with the advent of psychology, psychoanalysis, psychopathology, neurophysiology, anthropology, sociology and the narrative. The second notion perceives forgetfulness and oblivion as something that concerns all human experience and the whole of human history. It rejects partial horizons that originate in specific cultural fields (Freud or Proust or even Edelman) and instead turns towards metaphysics, the philosophy of history, the fate of Being and the West.

realisation outside time. Relying on time outside time removes oblivion from memory, replacing one for the other in a sort of permanent state of decline; serious experiences slowly die over time. Implicit (or procedural) memory is the paradoxical absence of these experiences.

In the twentieth century the subject of memory is focused on in fiction. A spatial object, a castle for example, embodies temporality serving as an instrument of narration. The *euphoria and stream of consciousness* in William James; the *objective correlative* of Eliot; Joyce's *epiphany*; Huxley's *counterpoint*; Lawrence's *fourth dimension*;Proust's *memory*; *the imaginary objectivity* of Gide;Valery's *myths* and *the labyrinth* of Borges, all incorporate traces of memory that contemporary society is unable to master. Fiction takes advantage of these oscillating phases between 'qualitative' time and 'quantitative' time; between individual time (biological and psychological) and collective time (solar, religious and political); between natural time and social time, slipping into temporal regularity and irregularity. Michail Bachtin (2001) borrows the term 'space-time' from Einstein to define a form of artistic interconnection whereby literature takes possession of single aspects of time and space from a historical or imaginative perspective. In 'space-time' the fusion of spatial and temporal features of a certain epoch stimulates the repossession of the past and the men who lived it. Time for the narrator of the novel is his own, inflexible to any other reference. Proust (1912) wrote:

> There is a geometrical programme and a spatial geometry. For me the novel is not only a geometrical programme, but it is also psychology in space and time. Time is this invisible substance I endeavour to isolate.

Proust's work moves in the past according to an idea that was in vogue at the beginning of the twentieth century: only the past is real and able to inspire art with its rehabilitation. This concept assigns to memory the capacity to evoke past experience, a chronological perspective[5].

According to this thesis, memory moves backwards and penetrates the past through a veil of oblivion, along sepulchral and sporadic paths, reconstructing meaningful evidence for the present. From Descartes to

---

[5] The paradigm of the re-emergence of the past through twentieth century thought characterises a great part of psychoanalysis and ethno-anthropological studies. For further reading see: E. J. Hobsbawm and T.O. Roger, ed. (1983), Cambridge University Press, Cambridge; A. La Vergata, (1990), L'equilibrio e la Guerra della natura: dalla teologia naturale al darwinism, Morano, Naples. (1991) and Nonostante Malthus: fecondità, popolazione e armonia della natura, 1700-1900, Bollati Boringhieri, Turin.

Damasio, the void that mysteriously divides the elementary material state from the elementary state of energy remains unchanged; for example, the nervous excitement caused by one's perception of the colour blue and one's sense of a very real interaction with blue. Unthinkable then as it is today, that unexpected emptiness one is faced with briefly appears only to vanish in the light that illuminates the moment. In the twentieth century novel this is the metaphysical moment in which the dramatisation of the action happens 'in the mind of the dreamer' (Deleuze and Guattari, 1983: 282).

Life can change in an 'instant'. There is nothing particular about the moment. One night you are sitting down to eat and without warning the life you have known is over. Everyone knows why. Each one of us has had first-hand experience – even on a small scale – of this instant. Just one word, one gesture; sitting at a table in a bar in a ray of sunlight. You were waiting for a caress, instead you received an 'adieu'. Usually people say'it just happened', or even 'it was so sudden'. Before, life went on as usual, after - nothing, finished. Sometimes you go into mourning, sometimes it only seems like a temporary absence. There is a tendency to believe that something went wrong when things do not turn out the way they could have, or how you would have liked them to. If things had taken a different turn, everything would have been wonderful, normal. But life is not like that. Things finish and that's that. The theme of absence and the pain caused by loss constitute the '*leitmotiv'* of metaphysics. According to Bergson (2007: 45-149), a fundamental repression marks the historical destiny of metaphysics; repression of time, duration, continuity, the present, their systematic 'substitution' with the *Ersatz* of material, the discontinuous, the composite. For Bergson, Heidegger and Husserl, the conclusive and hypertrophic knowledge our modernity calls 'psychology' is the testing ground for the hypothesis that theoretical history is the history of oblivion.[6]

To celebrate Freud's eightieth birthday in Vienna on 8th May, 1936, Thomas Mann (1980: 7) underscored that European culture had been ready for the Viennese scientist's concepts to take root and flourish – there was

[6] Heidegger, in his *Introduction to Metaphysics* (1977) was the first to discuss 'oblivion', saying that the history of metaphysics was the history of the oblivion of being. It was Aristotle who argued the 'theory' of the 'body', the 'being' supported by accident or predicate. Much later Descartes focused on the question of 'object,' *ego cogitans*, while Kant brought attention to 'preformed objectivity' from transcendental forms of cognitive subjectivity. Meanwhile Husserl (1954) reread the history of philosophy as if it were the history of humanity according to the theory of loss and 'recovery'.

'a feeling for illness, or more precisely, for illness as a means of knowledge'.

According to Freud, psychic illness revealed 'the underground world', which he associated with great human works of consciousness and art. The cultural environment in which Freud's formation matured is the very same space of antinomy which the masks of Greek tragedy highlighted. Man as creator and possessor of technology finds himself up against a 'double bind', in conflict with death and the gods (Gernet, 1983: 6). Death is an irremovable obstacle, and then there are the gods from whom men expect justice. Within this bind, Greek humanism accelerated its own contradiction, fostering structures to cultural history with the 'discontinuity of the present with respect to the past' (Sennett, 2000: 47).

The theatrical mask of Greek tragedy was the human mask. Its aesthetic role served to emphasise the distance between the two elements that occupied the tragic scene: on the one hand, the chorus, dressed in simple costumes, a collective character physically representing a group of citizens, while on the other hand, the tragic character interpreted by a professional actor, identified by the mask before an anonymous group, the chorus. The mask embodied one of those exceptional beings that legend had transmitted through heroic tradition. It represented one of the dimensions of fifth century Greek past – a distant past full of experience that contrasted with the order of the city but kept alive in popular religion in which the occult of the gods (ignored by Homer and Hesiod) enjoyed an important status. In the new version of tragic action, the hero ceased to be a model; he became a problem for himself and for others.

What happened between the sixth and seventh centuries in Greek cities constituted a new experience due to a series of circumstances that anywhere else might have been unproductive and without consequence. The Greek experience outlined a mental activity aimed at creating a space – one which was homogeneous, reversible, indefinite, reaching out in all directions whose reality consisted entirely of a group of characters and definable relationships. Being able to imagine such a space, and at the same time, to regularly practice within its confines, presupposed and prolonged a transformation, a description of experience and daily life. According to Vernant (1976: 39) the logic of this description is 'the logic of ambiguity, ambivalence, transformation or the inclusion of opposites'.

In fifth century Athens, the individual asserted himself as a 'subject with rights': the intention of the agent was recognised as the fundamental element of responsibility. However, neither the individual nor his relational life acquired sufficient consistency and autonomy to constitute an agent at the decisional heart of his actions. Action, however, did not mean so much the organisation and domination of time, but rather self-exclusion from time, a going beyond. Vernant (1976: 62) observes:

> Carried along in the flow of human life, action without the help of the gods is nothing but a vain, impotent illusion. It lacks the possession of that force of realisation, of that efficacy, the exclusive privilege of the divine.

Greek tragedy expressed this weakness of action, the agent's poverty, focusing on the gods at work rather than humans. The agent was no longer immersed in action. Being part of a temporal order the agent had no importance. He remained in the background, passively submitting to all without control of his actions. It was in this climate of impotence that the fifth century Greek challenged the incomprehensible gods. It was a daring challenge. However, in practice men were not up to what they had attempted to do. The evolution of the tragedy demonstrated their relative inconsistency, the lack of the agent's internal organisation.

Over a period lasting less than a century, the rise, the development and the decline of Greek tragedy signalled a strictly localised moment of crisis: there was a confrontation, at times painful, between the ever present antique forms of memory in the legendary traditions, and the new consciousness connected to the development of rights and political praxis.

Although the mode is different, the Promethean theme is relevant. Prometheus was certainly a symbol of human action, especially with regard to technical skills, but he was not a mortal. Mortals owed their tools of action to a superior being, their tools were subordinate to them. Protagoras, in spite of being a sophist, admitted Prometheus' hold; he was 'guilty' of having taught men technology transforming them 'from children into rational beings and masters of their own minds'. (Aeschylus, *Prometheus in Chains*). Prometheus claimed that he had invented numbers, the most important of all the sciences for men. Furthermore, he added, he had also taught men how to use letters, the memory of all things, the Mother of all arts.

For the Greeks, when artists or artisans produced a work with their *poiésis,* theywere not the real authors. Their role was simply to incorporate their *téchne* into matter with a pre-existent form. The perfection possessed in the work was superior to the skill of the craftsman(Vernant, 1970: 215).

'In the fifth century BCE, the Greek city set the pattern for the structure of philosophical reflection for the first time in human history, the boundary of adaptation was incorporated into the *méti;*[7], intelligence was

---

[7] As Homer's texts show, *métis* isthe specific intelligence which gets directly involved in practical difficulties rather than contemplating immutable essence, in

applied in the practice of solving difficulties using a strategy to achieve success in the most varied fields of action' (Detienne & Vernant, 1978: X).

The *métis* covered the cultural universe of the Greeks in all fields, from the oldest technical traditions to the organisation of the pantheon. In spite of the vastness of the field covered, in spite of the importance of its position in the scale of values, it was not clear what it was, nor was it apparent how enlightening its thought could be. It shone forth 'in all its profundity', immersed in a practice that was never concerned, not even when used, to make its nature explicit or to justify the way it acted. According to the philosophers, the expertise of *métis* meant skill, intuitive behaviour, sagacity, and farsightedness. It meant the capacity to transform difficulties into advantages, to remain vigilant, to sense opportunity, demonstrate ability in technological fields and acquired experience. But after many years the importance of the expertise of real knowledge diminished. Depending on the case at hand, it was dismissed and there was a reversion to unreliable opinion and downright chicanery. The *métis* continued tolive in those sectors where the philosopher usually passed through unnoticed or was spoken of with irony or in polemical tones, thus highlighting, by contrast, a different approach to reasoning and understanding one's own trade.

The 'word dialogue' is a 'privileged tool in social relationships' (Detienne, 1983: 77). It organised order in the Greek city. It was through the word that men participated within the assembly, commanded and exercised power over others. In dialogue, the word determined a closed space, a *logos,* an instrument of social relationships and a means of knowing reality. The unity of the mental universe that corresponded to *polis* was the abstract and interchangeable citizen – a *kosmos* in which men found their collocation, but not women, foreigners or slaves (Loraux, 1997: 13); a strictly ordered *kosmos,* but one of unstable equilibrium.

Such instability was the verification of two distinct ways in which Greek thought defined time – on the one hand there was *aion*, an ever present eternity, that had neither beginning nor end and therefore

---

spite of all the risks involved in the presence of a fickle universe of hostile and disconcerting powers. A *métis* god is always ready to jump into the fray. This does not mean that he loses heart, like many Homeric gods. On the contrary, his *métis* knows how to wait patiently for the right moment. *Métis* is quick, improvises according to the occasion, grabs the opportunity. Zeus is the god of *métis* par excellence. The Greek myths that tell of his conquest of power and founding of a lasting reign underline how victory after a struggle for sovereignty was obtained with *métis*. Zeus was not content to marry Métis the first time round - by swallowing her up, he became wholly *métis*.

perennially remained in the present and the future; on the other hand there was *chronos* with its quantifiable greatness, its temporal form of becoming and perishing, a continual transformation of the future into the past.[8]

Reference to the problem of time lay at the heart of the writings Aristotle dedicated to physics. Interest in time as in *aion* was replaced by a focus on time as in *chronos.*[9] The difficulty indicated by Plato in the confrontation between the eternal dominion of forms, the object of knowledge, and the reign of becoming of *chronos*, which for Plato is not available to knowing, found a solution in Aristotle. He attributed to time (as measurement of the process of becoming), the same characteristics that Plato and Parmenides assigned to the eternity of existence.[10] For Aristotle *chronos* was, in fact, *a-rythmos*, 'a number of movements that depend on what preceded and determine what followed', the act of becoming surged forth (an object of science) and consequently lost its *rythmos,* (the irreversibility of events), a qualitatively different process from a chance happening. Time was like a number mutationof *a-rythmos.* It inevitably resembled a geometric parameter of movement, a measurement of change, void of any effective dynamism. It transformed spatially contingent and intrinsically equivalent events into immutable permanence.

The indissoluble tension between this twofold concept of time represented the Western philosophical tradition that searched for eternal truth concealed behind the changeability of events, based on a vision of time 'without rhythm'. From this tension tragic theatre emerged in fifth century Athens, a frontier between memory and oblivion. The mask of Greek tragedy incorporated the 'disturbing extraneousness' of sensitive space in an attempt to render it absolute, preserving it from instability and destruction, in other words to liberate it from time and death. The ideal of

---

[8] These terms, alluding to two clearly different modalities and, to a certain extent opposing what is meant by time, appear in the first documents of archaic speculation, confirming the connection right from the beginning between problems of time and questions concerning nature.

[9] 'In Aristotle, *chronos,* even more clearly than in Plato, was and remains the subject of mathematics and physics, fully secularised and resolved in number' (Degani, 1961: 87).

[10] The scission between aionic time and *chronos* time, between the 'always existing' duration of being, and the irreversibility of becoming, is reconfirmed in the cosmological myth described in Plato's *Timaeus*. In this work, *aion* is the form of time referring to being, which therefore is shown to be eternal, while *chronos* is the time attributed to becoming, and like becoming is created and destined to die. Moreover, in Plato, *chronos* is not only distinct from *aion*, but is also an imperfect imitation of it, in that the becoming regulated by *chronic* time is the *mimesis* of being that has remained immobile in aionic eternity (Plato, *Timaeus*, 38-b). .

Greek death was this heroic desire to banish the horror of chaos and formlessness, to put a stop to the social continuity of humanity—modelled on the abstract space of the 'word-dialogue': human individuality had become a problem.

The cohesive bond between the social context in which the conflict of values appeared insoluble, and rigid praxis, an integral part of the world order, explains why tragedy came to the fore at a well-defined historical moment in space and time.[11] Greek tragedy used heroic legends for its themes. These deep-seated roots in the tradition of mythical stories showed that there was more religious archaism in great tragedies than in Homer. Tragedy expressed a detachment from heroic myths that inspired and transported its audience. It aroused doubt. It confronted heroic values and ancient legends with new processes that marked the demand for rights in the city. The tragic moment was when the distance in the social experience considerably increased due to clear delineation of the antitheses between juridical and political thought on the one hand, and mythical and heroic traditions on the other. But this moment was brief; the conflict remained as did the still painfully-felt conflict of values.

This debate between mythical past and urban present was expressed in a way that was particular to tragedy: man was focused on as an agent. In the famous carpet scene of Agamemnon by Aeschylus[12], the sovereign's

---

[11] Tragedy was born in Greece at the end of the sixth century. At the time of Aristotle's *Poetica*, written in the fourth century, the tragic mainspring had already been broken both in the public and in dramatists. There was no longer any need to debate the 'heroic' past, to confront the old with the new. Aristotle, who had elaborated a rational theory of action to distinguish levels of involvement of agents' acts, no longer understood the meaning of conscience and tragedy. For him they belonged to the past. The tragic effect did not lie in oneiric matter, as Freud (1900) believed, evoking Sophocles' Oedipus Rex to illustrate the 'Oedipus complex'. It lay in the way matter was shaped to make sense of the contradictions that lacerated the divine world, the social and political universe and the sphere of values. In so doing man appeared a sort of incomprehensible and disturbing monster, both as an agent and a passive subject, guilty and innocent, master over the whole of the natural world, incapable of self-government, aware but blinded by a delirium sent by the gods.

[12] Whether it is a question concerning impulse and desire as in Agamemnon, or reflection and premeditation as in Clytemnestra and Aegisthus, the ambiguity of the tragic decision remains the same. In either case the hero's decision corresponds to his personal character. Furthermore, in both cases it is evident that supernatural forces intervene at the heart of human life. The chorus has just reminded the audience of the cruel audacity of the King of the Greeks to sacrifice his daughter. They evoke the misfortunes of men as if they were responsible - 'the funereal

fatal decision undoubtedly depended on the baseness of human vanity. From the moment Agamemnon stepped onto the carpet, the drama was complete. According to Vernant and Vidal-Nacquet (1976: 28):

> the real tragic effect is derived from the intimate rapport together with the extraordinary distance between the banal act of solemnly walking onto a purple carpet with its very human motivations and the religious forces that are inevitably unleashed.

In this culminating point of the tragedy, the time of the gods appeared on the scene in human time (Vidal-Nacquet, 157, 1960: 55-80).The real sphere of the tragedy was collocated in this 'frontier zone' where human acts were articulated by divine powers revealing their true meaning. This meaning had previously been ignored by those who had taken initiative and become part of an order which had moved ahead of man, pushing him aside. There emerged from this 'frontier zone' a geography of scenic space behind which the inconsistencies of the director's inventions could be hidden. For example, in Cocteau's Orphée, the French director had a voice say off-screen that 'mirrors should reflect a little longer before giving us back our image'.

This voice tells the story of the tragedy of memory implicit in modern thought. It begins with a formula attributed to Simondi Ceo – 'the word is the image of reality'[13]. Implicit memory is the protagonist of the 'infinite dissolution'. Simmel (1977: 710) suggests that

> facts (action and social relationships), things (the world of things) and bodies (subjects, speaking in a phenomenological sense) tend to detach themselves from the layer of materialism in the world to produce

insanity that blows on the audacity of mortals' (Aeschylus, Agamemnon, 223) In the tragic decision the gods' designs and men's projects or passions collaborate together. Such a simultaneous presence seems to define the nature of tragedy with constant tension between the two opposite poles.

[13] Simonides born in Ceo in 557-6 BCE marks a turning point in the poetic tradition both for the type of man he introduced and for the concept of art for which he is the bearer. Above all, he is the first to conceive poetry as an occupation: to compose verses in exchange for money. He obliged his contemporaries to recognise the commercial value of the art and at the same time places the poetic function in a new light. Through painting, the poet discovered the *artificial* character of the poetic word. A complete tradition of the invention of mnemotechnic was attributed to Simonides (Vernant, 1965: 77, n. 98). Writing poetry as an occupation, defining the art of poetry as a work of illusion, making memory a mnemotechnic contributed to making 'word-dialogue' a turning point.

innovative ways between a movement of continual dissolution and becoming (Artieri, 2004: 18).

The Greeks converted a linear space into a symbolic space and projected it onto their stage. This changed the vectors of time from an active social process into the 'creator of its own causality' (Barba, 2009: 116) locked into a 'make-believe reality' (Stanislavski, 2000: 43), incorporated into the theatrical mask. Besides the mask, the culture of form confined the chains of motor acts that organised body techniques in the environment, 'the frontier of sciences where professors eat each other one by one' as Goethe says (I say eat but Goethe is not so polite) (Mauss, 1965: 384).

2.2 The Rhythmic function of "Body Technique"

Sociological interest in memory focuses on the combinative platform between genetic substratum and learning. Memory is not a unique process through which the subject autonomously conserves essential information. It is, on the contrary, an essentially relational phenomenon, and the formation of the mnemonic processes significantly responds to the distributed and situated character of knowledge.[14] During this formative process, more interactive dynamics are released by a synergy in which genetic reality and epigenetic development intertwine through the constant correlation between the biological dimension and the matrices of relational and socio-cultural order. The processes of learning and memory concern one of the most distinctive qualities of human behaviour – the capacity to acquire new skills by means of experience, and the ability to maintain this knowledge over time through the transmission of memory.[15] The capacity to codify and transmit memory was a decisive factor in the evolution of

[14] Today memory is conceived as a complex faculty of the mind that does not faithfully reconstruct the past but is an on-going process of production of recollections. Those recollections depend on meaning, sense and emotions, connected to lived experience as a social actor. This is particularly so when it is a question of important events related to the self in order to confer uniqueness to an individual's own life. Memory is in fact a multiple activity if we consider the diverse levels on which this continuous activity of selection occurs: each individual, and the community in general (families, friends, political and social groups, nations etc.) were particularly analysed by Halbwachs (1997) in relation to his conception of collective memory and society as an emerging reality today.

[15] For further reading see:

- M, Halbwachs, On Collective Memory, Chicago University Press, 1992.
- Jan Assmann, trans. Cultural Memory and Early Civilisation: Writing, Remembrance and Political Imagination, Cambridge University Press, 2011.

human society; its development was aided by the passing on of experiences acquired by their forbears to new generations.

This chapter focuses on the rhythmic function of 'body technique' elaborated by Marcel Mauss in 1936 (Journal of Psychology, XXXII, 3-4, 15 March-15 April). He believed that the body should be considered an instrument of synchronisation between the body's genetic programme and action between the topographic combinations achieved by each different technical phase. Such unrepeatable combinations transform implicit memory into working memory[16], progressively creating the source of active strength that forms a group's social memory and/or a society: objects for practical usage, artefacts, musical instruments, inhabitable spaces, repertories of song, music and dance transmitted from one generation to another. Collins (1990: 39) defines it as a 'chain of ritual interaction'. From prehistoric utensils in stone, bone or wood to the first ceramics, from machinery to the robots of today, these things have accompanied the human species on a long journey together. Changing slowly or suddenly with the times, places and systems of production depending on history and diverse traditions, they have, through the ages, always been attributed with new values and given a new aura of meaning.

As the Soviet scientist Simonov (2003) pointed out with regard to Stanislavskij's[17] acting technique, his was the affirmation of 'technique in process' in time, using the technique of his body. The neuroscientist compared and translated elements of the 'method of physical actions', step by step, into the scientific language of the day. Emotions were an electrochemical exchange between the nerve cells and the 'method of physical actions'; in his opinion they were functional to psychotherapy. Halfway through the twentieth century, Simonov's hypothesis was an attempt to move closer 'to the barriers between science and art' and becomes even more relevant when one considers how the concept of

---

[16] For further reading see:
- Richard Atkinson & Richard Shiffen, 'Human Memory: a proposed system and its control processes' in K. W. Spence & J. T. Spence (eds), The Psychology of Learning and Motivation, New York Academic Press, 1968, pp89-195.
- Elkhonen Goldberg, The Executive Brain: Frontal Lobes in a Complex World, New York, Oxford University Press, 2001, pp. 92-93

[17] Simonov, P.V. (1962), Metod, K. S., Stanislavskogo i fisiologija, Moscow. This work was partially reproduced and published in Santoro, L. A. (2003), work in progress, presentation and text notes, Lecce, Adriatica Editrice Salentina. The chapters that were translated and referred to in the text above are Chapters 3 & 4 concerning the theatre entitled *Method of Physical Action* and *On the Boundaries between Science and Art* respectively.

emotional adaptation today influences the most qualified analysis of these two fields (Damasio, 2007). Damasio (234) defines 'nuclear knowledge' as:

> a process that achieves a neural and mental configuration. The latter more or less instantly unites the forms relative to the object relative to both the organism and the relationship between the two. The emergence of these configurations and their confluence in time depends on the contributions of single cerebral sites that act in close collaboration.

In agreement with Damasio, it may be hypothesised that during the executive phase, body technique is naturally focused on the synchronisation of learned knowledge, and knowledge acquired through experience. The combination between the ability each individual contributes as a member of the *Homo sapiens* species including the ability to learn and remember, and the mutual exchanges characterising agent/context relationships need to be pointed out. What we incorporate into our individual mnemonic system is the function of our particular experiences; nevertheless, our existence and the basic function of mnemonic systems are the product of the genes of our own species. Body technique regulates this synchronism that temporally and spatially defines the immediate primary responses of the sensorial cortex from where experience originates. In the communication presented by Marcel Mauss to the Société de Psychologie on 17th May, 1934 he defined body technique as follows:

'I call technique an efficacious traditional act (and from this stand point you can see that it is not unlike a magic, religious, or symbolic act).What is needed is that it is traditional and efficacious. Neither technique nor transfer exists if there is no tradition. Above all it is in this that humans are distinct from animals – in the transfer of their techniques and probably more in the transfer of oral tradition.'

Mauss had no doubt that every act of learning was adaptive: learning presupposed a programme based on phylogenetically acquired information. This was what the notion of 'body techniques' demonstrated, once free of the Cartesian illusion of a fragile body compensated by a mind of 'eternal truth'. Another unconvincing explanation was that there was a pre-established harmony between organism and environment.

The categorical distinction between 'environment' and 'world' characterised modern philosophical reflection, and had considerable influence on a great part of German philosophical thought, including Heidegger. However, it was Arnold Gehlen (1967, 1983) who put forward the most coherent anthropological theory regarding the relationship between technique and human nature. According to Gehlen, man is characterised by a sum of morphological and biological peculiarities – premature birth, prolonged childhood, late endocrine functioning, organic primitivism, tactile hypersensitivity, innumerable motor coordination possibilities and so on – all of which contribute to situating him in a form of*exposition to the world* that has no equal in the animal world[18]. Subject to this exceptional burden, man is forced to find specific forms of exemptionin which technique becomes the paradigm. Over the following years, the concept that humans were the only animal 'radically open to the world' aroused legitimate criticism (if 'open to the world' is unilaterally interpreted as being a clear ontological demarcation between man and the non-human animal).

Such an interpretation came to be even more marked in the lessons that Heidegger gave on the ontology of life during those years. He made a very clear distinction between the 'disadvantaged animal', 'a prisoner of its environment' and 'man, the modeller of the world' (Heidegger, 1930). The idea of being 'exposed to the world' highlights the connection between technique and human nature and indirectly sums up the difference between ancient craftsmanship (techniques) and modern technological processes. The wealth of stimulation and the primary uncertainty of what being 'open to the world' means, constitutes a necessary premise for inventing new strategies. In other words, if a biological system creates its own equilibrium, starting with sound, for example, the latter becomes a precious potential to harness and actively produce[19].

From the flint chips of the Neolithic period to atomic power stations, genial technical resources respond to primary biological needs to dominate the environment, an intrinsic characteristic of human nature. In contrast, the arbitrary isolation of modern technique produces the myth of a possible original rapport with nature which in reality has never existed. No one doubts that modern technique is different from the past in methodology and efficacy, but this progress, no matter how amazing it might be, only confirms a cumulative capacity that has always been a normal characteristic of technological evolution. Moreover, throughout

---

[18] Without a set of biological filters that in every other species selects the decisive signals from the environment for survival, matching the most appropriate instinctive reactions, man is literally overwhelmingly challenged from birth 'by a spate of stimuli without any biological significance'; he is forced to make his way slowly 'without guidance through a chaos of stimuli' without any other natural gifts than his 'extraordinary willingness to react to perceptions deprived of congenital functions regarding signals (Gelhen, 1940: 62).

[19] Regarding sound as a positive resource and as a principle of order, see von Foerster (1984), *Observing Systems*, Intersystems Publications, Salinas, California.

history, society has reacted against what was considered abnormal technological innovation and not simply because it threatened consolidated habits. Humans are driven to transform their environment and adapt it to suit their needs, a necessity that is the same for many other species, if not for all species. Lewontin (1994:48) reminded us that every living organism is surrounded by an aureole of air heated by the body's metabolism. If the term 'environment' is taken literally to indicate the area surrounding an organism, then it is correct to say that every organism actively creates its own environment. It is a conceptual error to equate the environment with the 'objective' sum of data and the organism's external conditions.

When, at the end of the nineteenth century, Jacob von Uexkull (1956) introduced the concept of environment – *Umvelt* - into biology, his intentions were diametrically opposed. Uexkull's idea was that if we went to the country for the day it would be naive of us to think that the cricket, the cow, the ant and the farmer all shared the same space. In reality,the diverse instinctual functions and the specific motor sense apparatus work in such a way that the existence of each one of these organisms is unique in the world. One might be blind to colour or to shape while the organism living in close proximity has quite a different rapport with the environment and the world – a blade of grass could have a totally different biological function from one organism to another. As in the theory of systems, in biology the 'environment' does not indicate a specific objective fact, but rather the specific correlation of a biological system.

In Uexkull'smodel, the heart of the organism/environment correlation is made up of interactive processes involved in developing a generic programme characterising every type of organism. In other words every organism is genetically predisposed to react to a determined range of signals: the colour of flowers, the shadow of a predator and so on. Later, the concept of instinct defined above all by Lorenz (1974) confirmed this picture exactly. An instinct is, in fact, a specific motor sequence that is activated in response to a particular signal whose sequence of movements and the quality of the signal are genetically registered. Movement is not generated in response to a signal, unless it is in response to a predisposed programme. Every specific animal ability may be traced back to this 'network' between the genetically imposed programme and the action in context, including the particular constructive attitudes with which an animal actively intervenes in the external world.

Technique is the prototype of the human disposition to transform discordance into a new order. Every technical invention follows this transformation, revealing an effective weapon, for example a flint chip or mould, or a remedy against infection. An artisan's workshop exemplifies an active research site for solutions, a resource for survival. More precisely, the effective mastery of technique is to know how to make the most of every circumstance - Ulysses did not simply flee the storm but took advantage of the storm's strength to reach his goal with greater speed. This *autopoiesis capacity* exists in the rhythmic function of body technique and its ability to dominate over capital/time The body Mauss referred to concerns soma, emotion and behaviour, dimensions that shapeimplicit memory.That is a memory whose connection, notes Leroi-Gourhan, (1977, II: 233),

is strictly bound to the mastering of a mode of expression that cements man to the centre of the cosmos, but he still does not attempt to use reason and explore letters that transform thoughts into a penetrating line, fine as a thread, but long lasting.

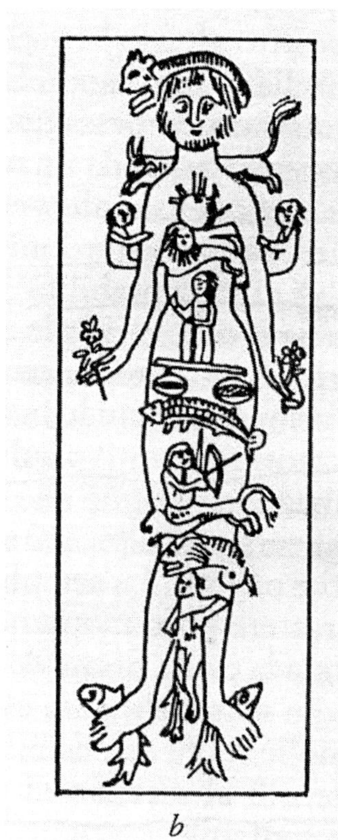

Figure 2-2 Source: Leroi- Gourhan, 1977, I, fig, 99b.

The living body undergoes a difficult and delicate encounter between physiology and activity. For Mauss it is a time loom that weaves the necessary base for community and society through the technique of the body, it embroiders the rhythmic passages of day and night, the seasons, the years and the generations. Body technique shapes activities of perception and motility, an unrepeatable combination from an evolutionary standpoint, between the oldest biologically acquired history of values, and the activity involved in successfully adapting to the environment. By 'body technique' Mauss means 'the ways man in diverse societies learns to use his body in the environment' (1968:365).

Body technique regulates an elementary system of accounting that varies throughout the planet, from region to region, transforming a memory aid into documents. Later, an even more rigorous accounting system replaced it. Only after the consolidation of urbanised agricultural organisms did documents reveal social complexity as serving as a testament for humans (or the gods). Mauss identified a means to connect biology and social relationships to the technique of the body, dividing the sexes by age, production and issue. The age of humans and their activity constituted the criterion for order that forged, according to Mauss, every single technique of the body. In this way he subdivided body technique into the technique of birth and the obstetrician, the technique of upbringing, feeding and weaning the child, the technique of adolescence and adulthood; the technical distinction of sleeping and waking. For every different technique he recorded times of movement and times of rest for different moments of the day. The interaction between action and temporal time delineated 'aspects of civilisation' that distinguished the social uses of the body. Mauss also outlined the techniques of rest (Mauss, 1965: 401-402):

> Rest can be total repose or simply a pause; you can rest in bed, sit or crouch etc. The way you sit is important. Humanity is divided by the way it sits – crouched or sitting, between those who sit on benches and those who use neither benches nor platforms, but who use chairs. A wooden chair holding a crouched person is very widespread – a most extraordinary thing – in regions 15° north of the equator. There are people who have tables and people who do not. The table is a long way from being universal. In the whole of the Orient people use a rug or a mat. All this is very complicated because during rest time people talk, they eat etc. Some people rest assuming most unusual positions. In the whole of the Nile region of Africa, and in parts of Ciad as far as Tanganika, the male population who works in the fields rests on one leg. A certain number manages to stand on one foot without any support, others lean on a stick. These resting techniques are

> authentic aspects of civilisation, common to a great number of individuals, entire families of a people.

Body technique practices have a temporal order based on the synchronisation of biological rhythms and the relational organisation of social space. In this 'rhythmic scale' biological rhythms are topographically ordered into practices that are vital for the individual and groups – nutrition, hygiene, work, relational behaviour and corporal *habitus.* Body technique creates conditions for knowledge that serve to build concrete social phenomena. Bourdieu (1980) demonstrated, in his systematic theory of practice, how knowledge, rather than being important for the mind in experiential contexts, is generated within these contexts during the course of reciprocal involvement with others in ordinary everyday life.

He suggested a practice of knowledge that is associated with ability, *know-how,* something we already have in our bodies that is difficult to codify in terms of systems regarding rules and representation. Consider the technique involved in lacing up one's shoes or breaking an egg or ironing. Activities like these develop not through formal instruction, but through repetitive performance, and often as silent tasks that involve a specific posture and gestures. Bourdieu called this corporeal *hexis*. The theory of practice suggests that perception depends on movement. An expert walker adapts his pace, monitoring it according to what he sees, hears and feels including the always variable conditions of the ground he is walking on.

In his study on *Body Techniques*, Mauss created a preliminary classification of posture and of the extraordinarily varied gestures used by people around the world in their everyday activity - resting (sleeping, crouching, sitting or standing) or moving (walking, running jumping climbing or swimming). Each body technique, concluded Mauss, is an 'acquired technique' (1936:390, 402-403). There is no 'natural way' of behaviour. He was concerned with how the biology of human nature was to be reconciled with each diverse technique of the body, as opposed to 'acquired ability' that models cultural expression. According to Mauss, the connection between human nature and culture might be established through yet a third term: rhythm. The social aspect of nature had been widely documented in studies by Wundt (1874; 1929)[20], Bucher (1896) and Ribot (1888; 1896), but Mauss (1965: 315) focused on the function of synchronisation between experiential contents and movement:

---

[20] cf. A. Oliviero Ferraris, and A. Oliviero, I motivi del comportamento umano, Zanichelli, Bologna, 2006.

> Let us leave the social nature of rhythm for a moment. From a sociological point of view, it seems obvious, for example, that even if one studies dance superficially it corresponds to respiration, cardiac and muscular movements which are identical in all individuals, even onlookers. This fact suggests a series of images, symbols of dance, which evoke that rhythmic beat in both dancers and bystander. Here, at this point, it must be understood that sociology and physiology are directly related and not limited simply to the social and psychological.

Considering passion as one of the effects of rhythm as well as song, and the way those who are caught up in it are passionate, there are numerous frightening ritual stereotypes that need mentioning: certain dances often accompanied by a simple shout, a scream to infinity, or a simple line in a song. Often, quite large groups are involved, searching together for excitement and ecstasy, day and night, through activity and physical fatigue. Cause and effect are simultaneous and reciprocal. Many happenings of this type occur in Australia and South America.

The physiology of all living beings is characterised by the repetitive functioning of tissue and cell organisms. Systematic study of biorhythms has demonstrated how the traditional concept of the homeostatic balance of organisms requires an urgent revision of the periodic variation of levels of activity, as well aswhat determines the best adaptation to environmental modifications. Many biorhythms, in fact, appear synchronised with natural or astronomical events, improving the use of potential energy for the organism. Body technique regulates the structure of synchronisations which are partly determined genetically and partly induced. Based on the variations of the external environment, the variations of the level of activity of the task at hand are monitored. The associative effect of body technique is focused on synchronising biorhythms and social rhythms. American physiologists, Sterman, McGinty and Adinolfi (1971, demonstrated several decisive recurring periods- for example meal times, bedtimes and the hour of waking - which constitute possible social synchronisations. Indeed, one of the most challenging frontiers regarding these networks is the biological base of the social network.[21]

---

[21] G.Bateson and H.R. Maturana studied human and social phenomena going back to the biological origins of our existence – not in the limited sense of biological psychiatry or social biology, but in biological ontological terms. They provided the social and behavioural sciences with the solid basis needed at the time. For further reading see:

- G. Bateson, Steps Towards An Ecology of Mind, New York, Ballantine, 1972.
- G. Bateson, Mind and Nature: a necessary unity, New York, Dutton, 1979.
- H: R: Maturana, '*Neurophysiology of cognition*' in P. Garvin (ed.) Cognition: a

The most relevant function of body technique is the production of arousal[22]. Such a function improves the performance capacity of the organism on a behavioural and physiological level. The state of arousal activates cognitive mechanisms – 'Be Careful!' At a conscious level participation is not required. The obvious developmental advantage associated with basic mechanisms that facilitate a quick evaluation of external objects and events explains the central role of these processes in our brain. In a millisecond the reply to 'Be Careful!' is 'Act!' (Davidson et al., 1990); a flow of energy is directed into the system through a process of arousal. Recent neurological experiments have demonstrated that the continual exercise of the state of arousal stimulates the neurogenic tissue. Only ten years ago, the first evidence of a real increase in cerebral structure related to environmental factors was achieved with the participation of London cab drivers[23].

The discovery is striking above all for its simplicity and explicative relevance. The hippocampus of London cab drivers is particularly large, larger than most people's hippocampus, since their work requires memorising a considerable quantity of complex routes and localities. Considering the importance of the hippocampus for memory, and given that London cabbies have to memorise a high number of routes and locations, their hippocampus is under great strain similar to the muscles of a weight lifter. In addition, the hippocampus is much larger in those who have been in this profession for a long time. Therefore it may be said that the dimensions of the hippocampus are directly proportionate to the number of years of activity. This suggests that there exists a direct rapport between the time of involvement in certain types of mental activities and the size of the neural structure involved in the activity itself.

The results obtained from the cab drivers are important from diverse points of view. Above all, it was observed how an important neural

---

multiple view, New York, Spartan, 1970.

- H: R: Maturana, '*Biology of cognition*' in H. R. Maturana & F. J. Varela, Autopoiesis and Cognition: the realization of the living, Boston, Reidel, 1980.
- H: R: Maturana, '*Cognitive Strategies*', in E. Morin & M. Palmarini Piatelli, (eds) L'Unité de l'Homme, Paris, Seuil, (1974).

[22] The state of arousal is manifested through activation of the central nervous system and the cortical component, thus increasing vigilance and improving performance capacity. For a comprehensive explanation see Izard(1992) and Ekman,(1992b).

[23] E:A: Maguire et al., 'Navigation related structural change in the hippocampus of taxi drivers'in *Proceedings of the National Academy of Sciences*, U:S:A:, 97 (8):4398-4403, 2000.

structure can continue to grow in a mature adult. What is even more revealing is the fact that the growth of a neural structure is stimulated by use. The effects of accentuated cognitive stimulation seem in this case to fare better than other effects of deterioration due to age. Despite the spectacular results obtained with the cab drivers, they constituted an exception. One research result is insufficient to draw conclusive generalisations.

Some years later, the reputable scientific journal *Nature* published a study based on magnetic resonance (MR) related to cerebral changes in jugglers[24]. Several healthy volunteers, none of whom had had any previous experience in the sector, underwent three months of intensive training for jugglers. By the end of the course participants were able to keep at least three balls in the air. When the cerebral magnetic resonances at the end of the course were compared with those carried out before initiating training, the results verified an increase in the grey substance of the temporal lobes for both hemispheres as well as in the parietal lobe of the left hemisphere. Such an effect diminished after the interruption of training and the advantages acquired in the temporal and parietal lobes decreased; a third MR examination after completion of the training demonstrated yet a further loss. From these results it is possible to observe the effects of the activity practiced on the proliferation of neurons in specific parts of the brain, even after a short time.

The notion of body technique shows, in synthesis, the variables that structure the function of the body in the process of constructing memory, and specifically in the construction of social memory:

1) The body as an integral component of the environment

In the development of thought, doing comes before understanding. No scientific description is equal to integration. Above all during the learning process, experience passes through the senses and actions are performed through movement. Body integration can never be substituted by a mere description. It is the ultimate descriptive source and the foundation of knowledge.

2) Synchronism between knowledge and know-how

Synchronism highlights the body whose spatial-temporal vector of distributed processes permits human beings through structural bonds in social networks to acquire, manipulate, recall and maintain information.

---

[24] B. Draganski et al., 'Neuroplasticity: changes in grey matter induced by training', in *Nature*, 427 (6972), pp.311-312.

Each new input (intuited by Hebb)[25], is integrated through new synaptic connections in a pre-existing network that is reactivated the moment information is added.

3) Rhythmic function

Insofar as theprocess is situated and distributed chronologically and geographically, body technique transforms biological rhythms into social rhythms. As Spanish neurophysiologist Joachin M. Feuster (1995; 1997; 1999), discovered, memory systems are built in concession networks, their formation reacts to elements of phylogenetically conserved experience in the phyletic memory of the primary sensorial cortex, making the associative cortex available for the development of individual and social memory.[26]

It is said that Erasmus of Rotterdam coined the following maxim: ‘you are not born a man, you become one’. Similarly, rhythmic versatility teaches us about body technique.

2.3 Biological Rhythms and Social Memory

The study of social memory[27] is one of the fundamental ways of approaching the questions of time and history; sometimes memory is

---

[25] Donald O. Hebb (1949), a Canadian psychologist, developed one of the most important theories on the formation of memory. According to Hebb, when one neurone stimulates another neurone, modifications in both their cells are produced; the arousal of the second cell increases the first cell’s efficiency. The principle of synchronic convergence states that when two cells are active at the same time they tend to join together in such a way that the activity of one cell facilitates the other. Hebb maintained that neurons, components of the nervous system, are organised into cellular assemblages that create prearranged circuits which gather and maintain information for a certain period of time. The fact that information circulates in the assemblages makes it possible to explain the process of memorisation and the ‘phase sequences’ that lead from simple to more complex behavioural patterns.

[26] In his analysis of mnemonic systems, Feuster recognised the primary sensorial cortex as a depository for a particular form of memory defined as phyletic, or a type of phylogenically determined memory containing adaptive experiences that the species has acquired during its interaction with the environment. The formation of individual memory is based on and developed on this patrimony inherited from the species.

[27] For further reading see:
-Niklas Luhmann, Theory of Society, Stanford University, Press,1997;
- Pierre Nora, Between Memory and History,1989;

situated in the past, sometimes in the future. In historical studies of memory, particular importance needs to be given to the differences between society and oral memory and society and written memory, paying careful attention to the transition period from orality to the written text[28]. Jack Goody defined the latter transition period in a publication entitled 'The Domestication of the Savage Mind' (1977).

All these theories referring to up-dating the more or less mechanical idea of mnemonic traces have been today abandoned for theories based on the more complex and specific activity of the brain and the nervous system. Changeux (1972:356) observed that 'the process of human memory not only includes the preparation of memorising but also the re-reading of what has been memorised'.

It is at this point that consideration for memory processes begins. Above all, their biological aspects are examined, showing that dynamic systems of organisation exist, in so far as they are maintained by dynamic systems and reconstructed by them (Assman, 2002, 2006).

The cognitive model of the human mind is based on the solidity of elaboration mechanisms. However, there is no network connecting the diverse processes, or an articulated theory that would unite different forms of cognition (or distinguish between them in a convincing way). Human Information Processing (HIP)[29]believed in the 'optimistic' notion of the existence of a unique mechanism of elaboration to apply to the entire sphere of thought processing which in a horizontal mode would link up and subordinate the connection of intelligence to the application of a single procedure.

According to Gardner (1987:299), there are two basic gaps in this model. On the one hand, HIP cognitive psychology presented itself as an a-biological and an anti-biological paradigm, analysing cognitive mechanisms without referring to the biological constitution of neural circuits from their origin. Instead, HIP uses the hardwareof the computer as a concrete base for the comprehension of mental states. Moreover, the analogy with both

---

-Josef Hayim Yerushalmi, Jewish History and Jewish memory, Washington University Press, 1982;

-Georges Duby, La Dimanche des Bouvines, 1973

- Luisa Passerini, Autobiography of a Generation, Italy, 1968, Wesleyan University Press, 1996, trans. Autoritratto di Gruppo, Firenze, Giunti-Astrea, 1988.

-M. Halbwachs, *On Collective Memory*, trans., L.A Coser, (ed.) Chicago University Press, 1992.

[28] Halbwachs, ibid.

[29] Neisser (1967) formulated the cognitive theory, defined as a 'synthesis model', on which *Human Information Processing* is based.

serial and digital 'intelligent machines' compresses the mind into a single cognitive dimension, ignoring those elements of creativity, spontaneity and emotion that are innate to it.

Post-behaviourism (Bruner, 1997b; Cohen & Segal, 1991; Striano, 2003) considers mental processes to be, on the contrary, transformative and adaptive resources. Defined according to specific contexts that guide their development and determine their function, they are based on internal relational evidence from where knowledge is distributed. At this point it may be said that the multi-systemicconcept of memory is a set of functional systems connected through converging zones.

Such a concept constitutes a most advanced theory regarding memory. Contributions from neuroscientists range from concepts of flexibility and plasticity to principles of localisation. This suggests the notion of body technique, where intelligences are regarded as expressions of a continuous process of transverse and synchronic exchange and communication.

In the formulation of the seven types of intelligences, recently extended to nine[30], Gardner (1987:228) includes corporeal/kinaesthetic intelligence characterised by two nuclei of specificity:

1. 'the capacity to use one's own body in differentiated and able ways for both expressive and concrete ends';
2. 'the capacity to work skilfully with objects using finger and hand motility as well as movements involving the whole body'

The strength of Gardner's theory lies in the fact that he attributes a value to the body which is not only sensorial, perceptive and physical, but is also a cognitive value *tout court.* He has gone beyond the typically Western tradition of culture, based on a 'radical disjunction between activities of the mind on the one hand, and physical activities, part of human nature, on the other.'

The Cartesian legacy is at the basis of a hierarchical process of human activity that values what is physical less than the symbolic, linguistic abstractions of thought. It is at this point that behaviourism tends to discern mental processes from physical ones, considering the former to be independent of, and parallel to, the latter but nonetheless superior.

[30] According to Gardner (1987, 1988, 1993, 1995, 1999) all human beings possess at least seven types of intelligence: linguistic, musical, logic-mathematical, spatial, corporeal/kinaesthetic, intrapersonal and interpersonal. Recently, he hypothesised another two types of intelligence – existential and naturalistic. See: 'How our intelligence is changing' – http://lgxserver.uniba.it/lei/rassegna/000417.htm.

Humans who have the slightest corporeal intelligence learn through direct contact with reality, exploring, manipulating, touching and interacting in continuation. Every individual relates to culture and the environment through a combination of corporeal/kinaesthetic intelligences and spatial intelligence in contact with contextual variations. The cooperative interaction between these different forms of ability not only guarantees the execution of complex human activities, but favours the development and the strengthening of the degree of care, efficacy and sophistication of their specific procedures.

Added to this ability, as evidenced in the toolkit of *Homo sapiens,* the practice of body techniques acquires a basic variable: the spatial determination of biological rhythms as vectors of adaptation. Body technique organises corporeal/kinaesthetic intelligence together with contextually collocated and relationally distributed rhythmic sequences, shaping the relational spaces between transactions of the 'innate' and the 'learned'. It is clearly evident that between 'innate' and 'learned' neither one dominates. Just as there is no single contribution of a gene, there is no unique contribution from the environment. Where there is learning there is a subordinate genetic mechanism. As Gary Marcus (2004:16) stipulates:

> The correct question to ask is not 'who' but 'how' genes work together with the environment to make a human being. Genes and behaviour design a map that situates the brain in its evolutionary context.

Through the exercise of body technique the temporal rhythms of life are gradually incorporated into the structural characteristics of things. Boas (1927:64) observed that:

> the rhythm of time appears to be translated into terms of space: the labour involved in breaking off flints, the use of the axe or hammer; the rotation of the wheel or the even pressure required by the potter and the loom. There is a necessary connection between the regularity of form and the rhythmic repetition of the same movement.

Boas clearly illustrated the composition of the processes to 'tame' the time needed by societies with oral memory. Oral memory is in fact a 'rhythmic body' of 'situated rationality' of circumstance (Boudon, 1987:2, 175, 203). It is limited to the exact dimensions of the situation and the occasion, and is exclusively identifiable in the concreteness of 'place and body' (Remotti, 1993), generated by body technique.

The technique of the body has a cultural correlation with the order of fluctuations[31], which could be defined as the order in which contemporary physics recognises the - self-organising time of biological systems, distant from any equilibrium. Diderot understood that control of nature was a laboratory of contrivance. A propos of inertia of the *système du monde* describedby Newton's laws of motion, in reference to the embryo Diderot opposed the dynamic complexity of real processes, defined by the location of the material in space and time (Diderot, 1935: 667), accusing this order of being unstable in appearance yet concrete. He objected to such generalisations used by classical science. The self-organisation of a given living system is in fact its capacity to start afresh as long as there is a distance separating it from equilibrium. 'Chemical clocks'[32] demonstrate, for example that dissipation of energy and material is, in conditions of instability, a source of order. Maturana and Varela (1980: 13)[33] defined self-organisation of a living system as *autopoiesis:*

> Autopoiesis in physical space characterises living systems because it determines the distinctions we can make in our interactions when we make them known. But we only know them when we can act with the dynamics of their internal states as composite units, or interact with them as simple units with which we interact in the environment.

It is a question of 'knowledge in action' that occurs within the limits defined by practical necessity. The basic sequences of the genetic code are

---

[31] Molecular biology discovered that 'non-linear' reactions are the rule in living systems. From the point of view of physics it is a question of exploring complex situations (very distant from ideal situations) that can describe classical or quantum physics in basic terms. Living systems on the one hand resemble a factory, a place of multiple chemical transformations while on the other hand, they present a remarkable spatial-temporal organisation with a distribution of highly varied biochemical material. Prigogine (1981: 171) described 'fluctuating order' thus: 'Far from being in equilibrium, fluctuations meant that we could use environmental differences to produce different structures. .Once more it is necessary to stress that the essential importance of conditions of distance from equilibrium - 'communication' and 'perception' - are the key words of the new comportment of material, distant from equilibrium.'

[32] 'Chemical clock' refers to a periodical chemical process which in experimental conditions is noticeable in a biological system that is at a distance from a state of equilibrium. See Winfree, Scientific American, vol CCXXX, 1974: 82-95.

[33] - H: R: Maturana, '*Biology of cognition*' in H. R. Maturana & F. J. Varela, Autopoiesis and Cognition: the realization of the living, Boston, Reidel, 1980.
-R. A. Cavallo, Systems Research Movement: Characteristics: Accomplishments and Current Developments, in General System Bulletin, vol. IX, n.3, 1979 :

incorporated into the rhythmic combinations that oral memory is based on, like a sonorous organism in movement. Oral tradition teaches us that thinking in mnemonic terms is to think in rhythmic terms. As one proceeds along the itineraries of sung words[34] one discovers in the music, the sound, the songs, the stories, the lullabies, the verses and the musical instruments, in the pharmacopoeia, a vast structure of ties in which, as in a concert, each living being and each thing are united in a specific place, reacting within a pre-established time. Every tie is a transaction generated by the exact intersection between the biological rhythms of the human body and the execution in progress of a specific body technique. Within this unique intersection, the muted body's place of action with its previous architecture falls apart; the elements that now exist make a selection from the variables of the mosaic combination of the day. Remotti (1993: 80) identified a particularly efficacious example of this procedure in the 'mobile capitals' of Africa. Under pre-colonial rule in Africa, on the death of a sovereign, the state institutions also died. The capital was abandoned. It was not possible to construct a spatial superimposition of two or more places of power. The eclipse of power generated a momentary state of disorder due to the competition between contenders for the vacant position. This momentary disorder is the matrix of the new order, expressed by the new sovereign and the new capital.

The notion of body technique elaborated by Mauss (1936: 391) may be defined as 'an adaptive itinerary pilot'. According to his demonstration, it is a temporal procedure to connect every specific phase of diverse biological rhythms with chemical flows and electromagnetic fields that at any given moment correlate sensorial and motor processes of the body with the context, generating the spatial functional 'combination' of adaptation. Basically, body technique regulates corporeal rhythms, orienting each phase within a specific composition of variable environments. Through such activity, human beings weave their own existence into the web of their ecological niche (Ingold, 2001: 29).

[34] In 1992, UNESCO created the programme Memory of the World in an attempt to counteract collective oblivion through the preservation of archives with collections of relevant texts for diverse cultures of the world that are the heritage of all but risk being lost (illegal trading, inadequate conservation practices, war etc). UNESCO has taken on the responsibility of safeguarding this heritage. The archive is organised according to geographical areas and nations but there are also inter- and transnational projects involved in the initiative.

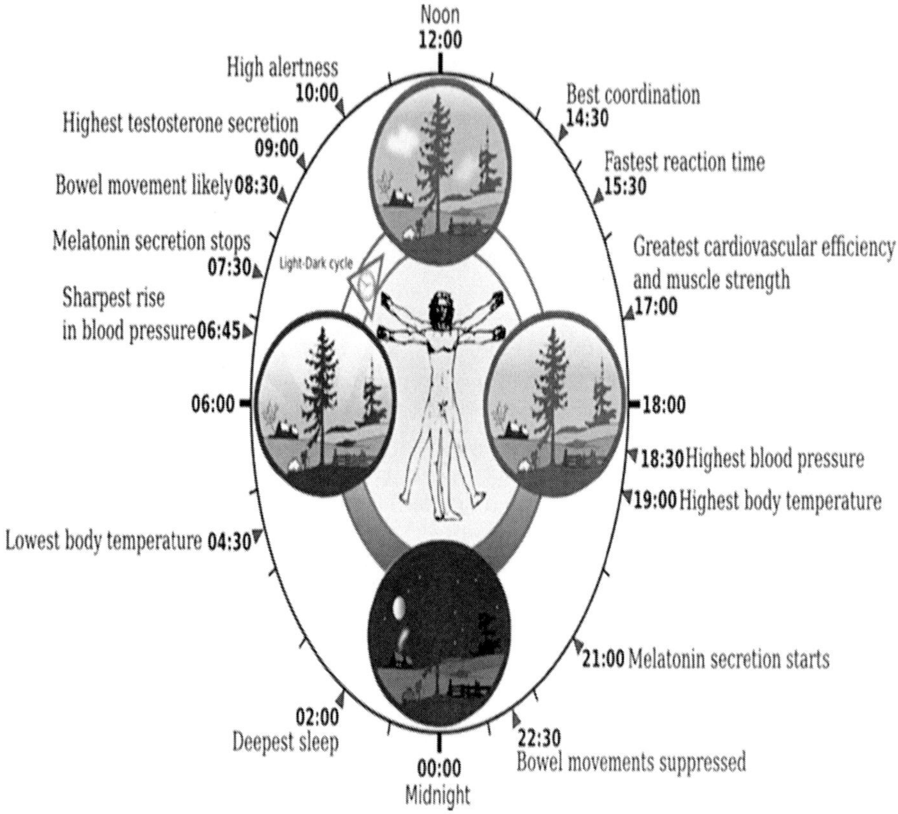

Figure 2-3: An overview of the human being's circadian biological clock with some physiological parameters from hptt://en.wikipedia.org/wiki/Chronobiology

Like all living organisms, the human body is also organised into a specific temporal structure in which all the vital functions vary over a period of time, ranging from a few milliseconds to many months. The existence of biological and specific emotional rhythms, in part genetically determined, in part induced by environmental influences, can explain the different somato-psychic reactivity of the surrounding environment. External influences are not simply superimposed, but are modulated by given rhythms from endogenous pacemakers. In defining the normal range of all the measurable physiological functions, period and volume values need to be specified including the phase relations of biological rhythms that concern the function. The principle biorhythms have a genetic base, having developed through an evolutionary process over millions of years. Even in human beings, as in less evolved organisms, many biological rhythms follow periodical stimuli originating in the exterior environment, while, conversely, many others are determined by internal pacemakers independent of environmental input.[35] For example, circadian rhythms synchronise the daily and nocturnal activity of the brain. Biological and emotional circadian rhythms are the most important and the most known since they correspond to a fundamental rhythm of rest/activity over a twenty four hour period (fig.3).

The systematic study of circadian rhythms shows how a great number of physiological functions and behavioural manifestations condition the individual to environmental demands throughout the day. With animals, there is usually a synchronism between periods of time and periods of activity (exploration, hunting etc.), while with humans such synchronism can be reduced or lost due to the complexity of the social structures within which they operate. There is, therefore, an excessive demand for psychophysical performance during a biorhythmic phase of rest, and *vice versa*. When the body is rested there is no complementary stimulation period of activity. Such de-synchronization of biological and social rhythms can lead to emotional and somatic disorders resulting from temporal stress.

In the antique Aramaic, Greek and Hebrew texts of the Old Testament, Marcel Jousse[36] demonstrated the close tie between body technique and biological rhythms, for example, breathing. Similarly, using the equation 'strength + matter = tools'. Leroi-Gourhan (1993: 224) explained culture as an 'exteriorization object'. Body technique adds two fundamental variables:

1. rhythmic rule - action based on exercise of biological rhythms;
2. spatial progress as defined and orientated within a specific context of action.

The intersection between these variables synchronises memory's heritage, becoming visible at the opportune moment using elementary topography

---

[35] Organisms acquire a biorhythmic comportment taking advantage of day and night, waking hours and hours of rest. They regulate nutritional and reproductive activity according to the seasons, migrating to maintain more suitable climatic conditions. Biorhythms influenced by the environment are called 'exogenous', rhythms determined by physical disturbances such as cardiac conditions or pulmonary activity are called 'endogenous'.

[36] Marcel Jousse, 1981, (trans.), Rhythmic & Mnemotechnical Oral Style Among the Verb-Motors, Fondation Marcel Jousse, Diffusion Le Centurion.

to determine practical action. Invent, dispatch, payment of salaries, calculations of profit and loss, and reports all constitute practical support for *commercium,* handed down from generation to generation between the human body and the rhythms of living matter.

The system of seals from protohistoric India constitutes a particularly significant example of memory heritage[37]. As far back as 3,000 million years ago it connected urban communities of different ethnic, linguistic and cultural backgrounds over a vast area of the Indus valley, extending from the mountainous valleys in the north to the plains in the south, from the deserts to the steppes, from the banks of rivers to the swamps and ocean shores. Incomplete seals revealed that there was a precise technique in their fabrication. The artisan engraved the seal in separate parts, completing for example the head or the body of an animal before outlining the general proportions. The surface of every seal constituted a small square in steatite carefully sanded with a perforation on the back. (fig. 4).

Horned tigers, unicorns with strange trunks, cattle with three heads are some of the designs depicted on the seals, according to technical codes that are well-defined and repetitive, at the same time that each one differs in the way it treats the artisans' repertory, handed down from generation to generation. The technique of the seals reflects the power of Harappa society, the hierarchy of settlements, the system of measurement and administrative practices. The absence of temples and large royal tombs indicates how the organisation of power developed in these societies, as political and economic integration between relatively independent communities. This integration was probably centred on the distribution and exchange of precious materials such as gold leaf, jade, alabaster or cornelian, probably traded with Iran and Mesopotamia.

On a par with the Harrappa seals, the system of weights and measures afferent to the Near East was a procedure for registering economic transactions, as archaeologists have hypothesised. (Amiet, in Archeologia, XII, 1966:20-22; Schmandt-Besseratin Syro-MesopotamianStudies, I-II, 1977).

The topographic combination of the repertory of memory within the weft of physiological time limits, produced by the exercise of body techniques, organised the social memory of Palaeolithic culture, and great Neolithic civilisations which developed in fluvial basins and in cultures of primary orality. Leroi-Gourhan (1977, II: 331) observed:

---

[37] In the absence of arms or military structures, stamped seals were used to denote individual rights of access to territory belonging to different social groups guaranteeing systems of alliances and parentage.

Alternating between sleep and waking, digestion and appetite, all physiological time limits form a weft within which all activity is incorporated. These rhythms are generally linked to a larger weft, the one that alternates day and night, the one that regulates meteorological changes and the seasons.

Figure 2-4: A seal in steatite, Harappa Museum

Until the dawn of printing, the greater part of knowledge remained in the hands of technology. Innovation could be found in financial and religious acts, genealogies and calendars. As the perfection of instruments for transferring generations of words, phrases and sentences to memory progressed, whole areas of knowledge destined for future generations were limited to selected fields by the hierarchy of social values. The diffusion of printing increasingly aided the *exteriorisation* of individual memory while dictionaries became a fairly advanced form of external memory. At that time the art of documentation was at the same level as mechanical animation; the highest point of the automaton was animation through separate cams, which conserved a localized memory for every organ. In modern times, biological rhythm is limited to modelling the corporeal 'periphery' as if it were a concealed world that has never been written down, that no one has ever known but everyone understands (Sapir, 1949). It is what Hall (1968) defined as the 'hidden dimension', referring to 'profound, shared and imprecise experiences' that members of a particular culture have in common: unconsciously communicated, they constitute a background against which other events are judged. Such experiences are generated by deeply rooted biological rhythms in diverse *habitats,* in which the body moves and interacts - the language of corporal posture, the way of walking, the organisation of culinary habits, family relationships, ways of measuring time and space. Body technique activates them all, depositing its own findings into implicit or procedural memory.

The idea of space derived from Newton's experiments, and the way people thought of it in the seventeenth century, meant that real places in the world were abandoned, to the detriment of the geography of place and body time, assigned instead to a latent memory. Bergson discovered this profound, personal and 'pure' memory hidden behind a superficial memory that was anonymous and comparable to habit. Even Surrealism, modelled by dreaming, was obliged to take memory into consideration. As far back as 1922, in his *Carnets,* André Breton queried whether memory was nothing but a product of the imagination. In order to know more about dreams, man must put his trust in memory, no matter how fragile and misleading it might be. From this derives the importance that the theory of 'trainable memory', a new metamorphosis of *artes memoriae,* had in the *Manifesto du Surréalisme* (1924). At this stage Freud's influence must be mentioned. The *Interpretation of Dreams* affirmed that 'the behaviour of memory while dreaming has without a doubtan enormous importance for every theory about memory' (Freud, 1967, III: 28). In chapter two, Freud deals with 'memory in dreams'. Taking an expression from Scholz, he sustained that 'once something has been intellectually possessed, it cannot be lost'. He was critical of the tendency to 'reduce the phenomenon of dreaming to a genre of remembrance', since there is a specific choice of the dream in memory and a specific memory of the dream. The memory in this case is chosen. However, Freud was not tempted at this point to treat memory as if it were a thing, such as a large reservoir. Instead he collocated the dream to latent memory and not to conscious memory, insisting on the importance of childhood in its formation, thus contributing, along with Bergson, to the exploration of the sphere of memory, and highlighting - as far as individual memory was concerned - the tendency to censure memory, which is so important in manifestations of collective memory.

In 1894, when Freud began investigating the role of unconscious mental processes in behaviour, he was trying to carry out empirical psychology at the same time. Due to inadequacies in neurological science, he put aside the biological model in favour of a model based on the verbal confirmation of subjective experiences.

Today it is known that the synchronism between the elementary consciousness that regulates somatic processes and mental/corporeal interaction creates the unrepeatable activation that produces active consciousness. The body is the cornerstone of this synchronism. The notion of body technique introduces a consideration of the body as a material agent of social practices directed towards the cohesion of the group, capable of improving the management of those relationships that depend on short- and long term wellbeing.

2.4 Why Dance?

The original place of dance is the human body, in so far as it is the crossroad between heritage and learning. The body is in fact the result of the unrepeatable coming together of environmental components, casual events and hereditary traits. They shape the dynamic function in the development of the mind through learning processes. The bond between body and dance is such that both man and the environment are reciprocally changing with their activity[38]. John Blacking, the ethnomusicologist (1973: VI-VII) made the following observation:

> The fundamental principle for most, if not for all, is that the essential processes of music are to be found in the human body and in the systems of social interaction of human bodies.

Dance is the spatial picture of which actions are a part; positions suitable forlearning and playing are established, and similarly, where to meet and where to hide. Dance coordinates the activity of the group and it holds it

---

[38] For some time the term 'environment' has indicated a series of essentially natural processes, considered to be at the origin of all that surrounds man. The European etymology for the term is basically the same for all languages of Indo-European origin: 'en' (around) and virer (turn), a term that is absolutely *anthropocentric*. Man is not seen as an integral part of the biosphere, but rather as an entity and a factor outside it capable of modelling and managing the 'environment created for his own needs and in virtue of the superiority of his intellectual gifts'. Such an anthropocentric vision (already present in the Greco-Roman world and valued by Western Christian thought) was slightly modified after the second industrial revolution, and more particularly during the period after the Second World War. Demographic growth, the strong impact of technological development, greater scientific knowledge and the diffusion of mass opinion have promoted a greater awareness of the risks man is running, and a less anthropocentric perception of the environment Generally speaking, the concept 'environment' can be divided into three main notions:

1. Spatial: enclosed space, habitation, edifice where one lives or carries out an activity;
2. Figurative: social, cultural, historical, geographical, conditions in which one lives;
3. Scientific: several disciplines - and in particular ecology - that are the object of analysis.

together, keeping humans at a certain distance of communication so that they can inform each other as to the presence of food or enemies. The crucial nature of dance is, in short, the *rhythmic construction of territoriality.*

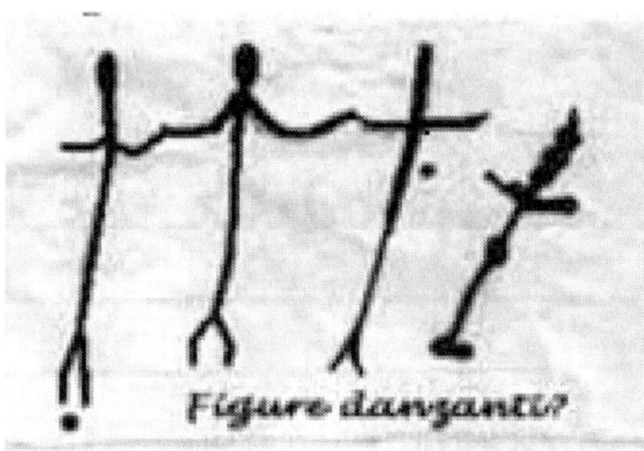

Figure 2-5: Neolithic rock carving on a block of stone about 30 metres square on which 30 figures have been carved and 7 basins excavated, at Ciappo delle Conche, an archaeological site near Finale Ligure, close to Orco Feglino, at the southern foot of Monte Cucco.

In modern culture, dance is a typical transversal term, half way between two large categories of philosophical reflection: space and time. Interest in the problematic nature of space is inherent to its very concept. One of the first elements related to this complexity is due to the experience of distance between one point in space and another. To cover this distance requires time, and it is precisely this combination that gives rise to the problematic relationship between space and time: how man physically moves in space, and what forms and solutions and diverse social constructions determined by societies and historical contexts are involved. The distance between bodies is a repetitive element that human and social conditions have in common. Just as the construction of social aggregations based on spatial proximity is taken for granted – from the village to the city – the way of life, sharing limited spatial territory, cannot avoid face-to-face contact in collective traditions.

In his analysis of the spatial behaviour of living beings, Hall (1959; 1963, vol. 65:5; 1964, vo. 66: 6; 1974; 1976; 1983) demonstrated thatman's limits coincide with his own body. For example, the presence or absence of the sensation of another person's body heat marks the limit between intimate and non-intimate spaces. According to Hall, territory is a very real extension of the organism, outlined by olfactory, visual and vocal signs. Man has forged new extensions of territory and with it new visible and invisible territorial limits. One of the most important functions of territoriality is each individual's own space, the preservation of that part of one's habitat every species depends on for survival. Territoriality is also associated with position. A series of experiments carried out by Bain (1949), a British ornithologist, demonstrated how power relations concerning 'feeding station' positions are modified or even substituted, among birds living in adjacent territories. Man, too, has his own territory and has invented many ways of defending it – considering it 'his land', 'his field' or 'space regarding his competence'. Perhaps the most effective of these territories is dance. Dance outlines and orientates the pathway to learning through the scores of actions that make up every diverse sequence adaptation. The information stored in the nervous system as a result ofthe experiences each organism undergoes while interacting with its environment determines the type of more or less permanent adaptive modification of behaviour usually called 'learning'.

Let us take the better known example of bees in the animal world, comparing their characteristics with those of humans. On returning to the hive after having deposited the nectar, the forager bees do a vertical dance in the hive. If the food is located within 50 to 100 metres of the vicinity, each forager does a circular dance turning first to the right and then to the left for approximately one minute. If on the other hand, the food is located 100 metres further away, they do the so-called waggle dance – they move forward a short distance in a straight line, rapidly moving their abdomen, followed by a 360° turn after which they repeat the movement forward in a straight line followed by the 360° turn, but in the opposite direction. The repetition of these moves creates a sort of figure eight indicating where the honey is located. (fig. 6).

The distance the forager dances wagging its abdomen indicates the distance from the food source – the further the food, the more the forager waggles its abdomen at a regular speed, completing the figure 8. As far as the direction is concerned, the angle of the straight line of the wagging of the abdomen with respect to the vertical line corresponds directly to the angle calculated from the hive between the direction where the food is located and the direction of the sun. Gould (1984) demonstrated that the foragers were willing to leave the hive after having assisted in a dance indicating the location of the food on the other side of a lake, but not when the food had been located in the middle of the lake. This reveals a surprising degree of integration between communicative capacity and spatial knowledge of the territory.

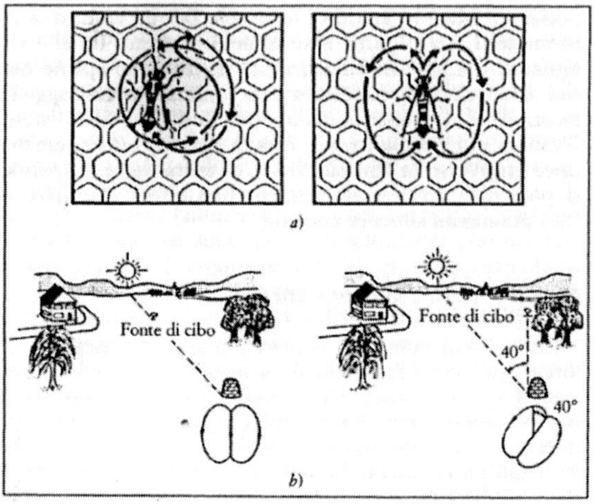

Figure 2-6: The honeybee dance: in diagram a) the circular dance (on the left) and the waggle dance (on the right); in b) how the waggle dance indicates the source of food. The angle between the orientation of the dance and the right angle is the same as between the source of food and the sun, measured with respect to the entrance to the hive. (Source: Pearce, 1997).

On a par with animal communication, territoriality and learning also describe the origin of dance between humans. Through relationships, human beings have a way of imposing behaviour and values that contribute not only to their short term wellbeing but also to their biological interest. Numerous studies indicate that the development of the brain can be regarded as the product of the effects that experiences exercise on the expression of genetic potential (Kendler & Evens, 1986; Cohen et al., 1992; Gunnar, 1992; Rakic *et al*., 1994; Goldsmith *etal*., 1997; Rutter *et al.,* 1997). Genes contain the information that stimulates neurone growth, uniting or dying during the processes that differentiate cerebral circuits. These processes are therefore genetically programmed but they are not 'experience-dependent'. Changes in the processes that regulate the organisation of cerebral activities – emotions, or long term memory – are linked to structural modifications of nervous circuits, caused by the activation or deactivation of genes which codify the synthesis of specific proteins. In this picture the genetic expression, the activity of the mind, the behaviour of the individual and his continual interaction with the environment (experiences), are all closely connected parts ofa single process of development (Rutter, *et al.*1991).

Both music and language are linked to movement. Moreover, to understand their evolution we must consider the entire human anatomy. It is possible that the effects of bipedism on the way humans move and use their bodies as a whole – the brain, the sensorial motor and the vocal apparatus – marked the beginning of the greatest rhythmic revolution ever produced during the course of human history – dance. *Homo ergaster*[39] walked erect on two feet exactly like *Homo sapiens.* It is this evolutionary development – bipedism – that we must take note of in that it had radical implications for the evolution of our cognitive and musical capacities At first it might seem surprising that their origin principally lies in the evolution of the brain, the vocal apparatus and their social life. The social complexity of ancient hominids and the first humans created selective pressures for a more developed communication. Physiological changes, especially those associated with bipedism, meant that *Homo Ergaster* could communicate using a significantly better system than the one available to his immediate predecessors – hominids and modern anthropomorphic monkeys. The new level of motor control, the independence of the trunk, the arms and the legs, and the rhythmic capacity considerably increased the body's potential over pre-existent possibilities. As Leslie Ajello of the anthropology department at the University College of London explained at a conference in 1996, the origin of bipedism had deep implications for the evolution of intelligence and language. The fact that *Homo ergaster* had a larger brain than the Australopithecine was probably because of a new level of sensor motor control that bipedism required and at the same time accepted. The erect stance and bipedism require that the centre of gravity be constantly monitored and that small groups of muscles alternate frequently to correct the stance, while leg movement must be integrated with the arms, hands and trunk to maintain a dynamic balance. Furthermore, bipedism requires a larger brain and more articulated nervous system to operate this more complex sensor motor control. Once developed, this larger brain was able to programme other tasks, such as procuring food, social interaction and

---

[39] Today, the most convincing argument regarding the origin of bipedism is that there were two distinct stages in its development, each one characterised by selection pressure. The first brought about the partial bipedism of the Australopithecine and the second led to the complete bipedism of *Homo ergaster.* Regarding the first phase, Kevin Hunt of Indiana University suggested that fruit picking had a crucial role to play, after having studied chimpanzees and observing how they used their hands to move from one tree to the next, saving energy by not having to continually raise and lower the body's barycentre.

language. Originally Ajello argued that the *fulcrum* of the process was the evolution of the hominid body rather than the brain.

There is no doubt that body means movement – also and above all for *Homo sapiens*. From the making of the first hunting tools to the building of a cathedral, from an elementary gesture or sonorous communication to the advent of optic fibre, corporal motility constitutes a *continuum*. Every motor act is the emergence of a new function, prepared by the previous phases in a sort of self-structuring, compared to the environment. The motor act is at the basis of human movement, and to obtain it a correct relationship with the environment is necessary. In order to interact with the external environment, the body must be able to recognise the situation and foresee the variables – a task delegated to the sense neurons and their receptors. Kandel (1988: 330) upholds the theory that afferent fibres transform physical energy into nervous information while efferent fibres transform nervous information into physical energy. The motor systems always receive external information through the sensorial system: vision receptors, distance perception and the muscular and vestibule state thus making it possible to define the position of the body in space, permitting it to carry out correct movements consuming as little energy as possible while foreseeing and providing the modalities for adaptation as needed.

Movement is founded on kinaesthetic time.Diverse fromrhythmic time, kinaesthetic time is articulated by the movement itself as a unit of measure, and its duration is given by the minimum quantity of real time necessary to complete the task with the maximum of efficiency. For example, athletics can be distinguished from other forms of movement because it is organised exclusively in kinaesthetic time. To correspond with acceptable requisites, athletic movement must only last long enough for the action to be carried out with the maximum of efficiency. Acrobatics constitute another example which clearly demonstrates the difference between kinaesthetic time and rhythmic time. If the acrobat's movement is programmed to reach maximum efficiency in minimum real time, the execution is carried out exclusively on the *continuum* of kinaesthetic time and is considered a form of athletics. On the other hand, if the acrobatic movement is carried out according to rhythmic time, then we are talking about dance. At the Olympic Games a gymnast who performs on the parallel bars with skill is practising a form of athletics while the same gymnast performing the same exercises but on the ground, accompanied by music, is performing a type of dance.

Even though dance is categorised as movement, 'it is more than just a part of the movement' as the Calouste Gulbenkian Foundation (1980) pointed out.

As the analyses of ethnomusicologists and anthropologists have also highlighted, the choreographic repertory achieves, in a specific dance movement, an itinerate order between human action and environmental matter intent on correlating the mental development of the individual with the experience of his body, a progressive task computed in the environment. For example, in discussing the social function of Indian Kwakiut dance in the north west of North America, Franz Boas (1981: 17) observed that:

> Everything associated with the life of the Kwakiut is accompanied by some form of dance, from the cradle to the grave. When the newly born child is resting in its cradle hanging from a branch, the mother ties a cord from the branch to her finger or toe and rocks the cradle backwards and forwards singing a lullaby. When the child is a little older and is able to stand, the father holds him over his knees and to the rhythm of a ditty, the child 'dances' and the father sings:
>
> *When I grow up I'll be a hunter, Oh,dad, da, da*
> *When I grow up, I'll be a climber, O dad, da, da*
> *When I grow up, I'll build canoes, Oh, dad, da, da*
> *When I grow up, I'll be a lumberman, Oh dad, da, da*
> *When I grow up, I'll be a labourer, Oh dad, da, da*
> *You won't need a thing, Oh dad, da,da.'*

Since it is important not to make mistakes in beating the time, not all those who sing are allowed to beat the rhythm when an expert dancer dances to a tune with a complex beat.

Beating time varies from culture to culture, from society to society. African dancers utilise time with an impressionable and sophisticated use of spatial elements. All African dances take place in the open air on ground, often extensive, reserved exclusively for the purpose. The drummers who accompany the dancers and who are an integral part of the performance, occupy a pre-established position – in some cases central,but often on the outskirts of the area, as far away as possible from the spectators. African dances are always oriented towards the drummers, and the pericorporeal space and the movement through space are made use of to the maximum, with great attention and sophistication. In the region of West Africa, dance always plays an important role, but sometimes 'an extraordinary role' (Gorer, 1981: 49) in the social, religious and aesthetic life of the community.

As the above-mentioned examples testify, the primary function of dance is to synchronise body rhythms with the chronospatial maps that model the spaces of action of each diverse biological rhythm, reviving the memory networks that over time progressively print the history of territorial adaptation into the human body. In the preface to the History of Dance by Kurt Sachs (1981), Diego Carpitella observed that:

> The continual comparison of the human body with the immutability of limits in the world environment seems to confirm that body techniques are among the most resistant of the human species and if it is true that the circle of the Bororos is different from the folkloric Cordella because of the difference in context, steps and movements, clarified by choreographic procedures, then the continuity established between the human body and the space it determines, between the configurations of corporal instruments and the cultural phase is also true.

Rhythm is as fundamental to music as are melody and harmony. It is all about the length of time between the sounds that follow each other[40], one after the other; it is about modelling the body's movement within the interactive context[41]. George Leonard (1981) who has made a study of rhythm is convinced that nothing happens between human beings unless it corresponds to a rhythmic hierarchy. Like Leonard, John Dewey was also interested in rhythm. In *Art and Experience* (1959) he demonstrated how rhythm pervades all artistic creativity: painting, sculpture, architecture, music and dance.

Dance regulates, at least in its initial stages, the sequences of biological rhythms through the contextual exercise of primary emotion,[42] 'the oriented response' of the 'body in action' in the surrounding environment. Siegel (2001: 124) observed *à propos* of emotional processing:

---

[40] The term 'rhythm' is here used in the sense of temporal processes referring to different sequences of human behaviour, and more specifically to micro-time, which gives structure to the limits of interactive synchrony.

[41] Context means forces, phenomena and tendencies which can have an environmental, economic, political and social nature that conditions and influences the choices and behaviour of an organisation regardless of the actors involved in the system.

[42] According to Siegel (2001: 125), 'primary emotions' are cerebral states generated by oriented responses, *arousal* processes and elaborated valuations. They are distinct from the concept 'basic,' or 'fundamental discrete emotions' which refer to differentiated mind states. Non-verbal, and often unconscious, emotion induces a state of vigilance in behaviour. This definition of primary emotion is in accordance with Ross Thompson's (1994) 'emotional dynamics' concept that involves immediacy, intensity and specificity of emotional responses. A systematic analysis of the conditions in which these complex and integrated modifications are manifested demonstrates that they appear in circumstances that involve the survival of the individual and the survival of the species.

> This reaction activates cognitive mechanisms (Be careful!) that do not require participation at a conscious level and initially are neither positive nor negative (Ledoux, 1990; Barbas, 1995). The brain very quickly processes representations (we are talking about microseconds) referring to the functions of the organism and to the external environment at a particular moment. At this point a phase that we might define as 'elaborative valuation and arousal' takes over. It is characterised by processes that model the flows of energy to the brain through the activation of certain circuits and the deactivation of others which differentiate an individual's states of mind (Barbas, 1995; Rolls, 1995; Lewis, 1996).

This "response" models the emotional processing in learning. Action is at the basis of learning (Rizzolatti, 2006) and there is not only one way of learning. There is an archaic mechanism that is found in animals as well as humans. As is seen in children, understanding precedes language. Jean-Pierre Changeux and Paul Ricoeur (1998: 162) emphasised that potential motor acts[43] that dominate, like 'orienting and hunting activities' or 'sequence motor interventions', depict the world as a 'workable environment'. Such a world does not only depend on whether we take this or that object (or whether we are ready to do so), but rather our capacityto move and find our bearings in the space that surrounds us, let alone understand the movements and actions of others. This is what dancing bodies in a circle reveal about human beings and the carved animals in the central cave of the Addaura complex in Mount Pellegrino near Palermo (fig. 2-7). [44]

---

[43] Movement always needs an immediate 'where to' – a direction, or feedback. At birth, when the central nervous system is still immature, apparently chaotic motility is waiting to be given some order from the learning process that biological growth brings about. The human neurological organisation is a two-sided game between inheritance and environmental functions. All individuals belong to a *genotype* and a *phenotype*. Where the *genotype* is all that we were, what we are and what we will be, *the phenotype* arises from the individual strategies that can play a role inside the forces of gravity in the environment.

[44] Bones and tools used for hunting have been found in these caves giving evidence of human settlement from the Palaeolithic and Mesolithic periods. These relics may be seen in the Archaeological Museum of Palermo.

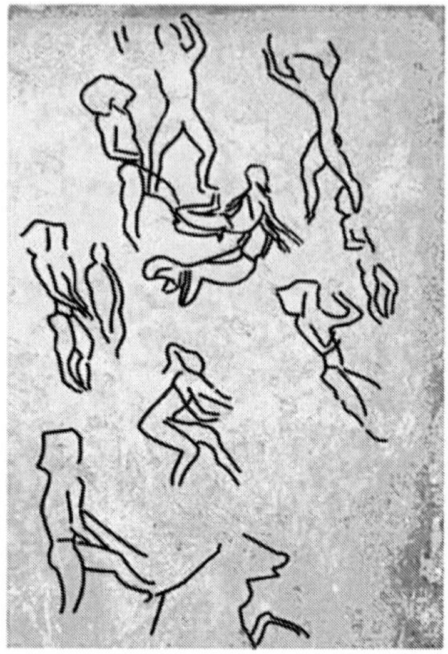

Figure 2-7: Wall etchings of dancing human bodies and animals in a circle, dating back to the Epigravettian and Mesolithic periods in the Addaura, caves, Palermo, Italy.

The organisation of the chain of movement that shapes all dance choreography outlines the territory of action based on sensorial interaction, extending the physiology of the biological body to its cultural dimension through music. John Blacking (1975) the ethnomusicologist and social anthropologist wrote:

> It seems to me that basically, what is more important in music is that it cannot be learnt in the same way as other cultural practices are learned. It is a question of something that is already inside the body, waiting to be expressed and developed, in the same way as the basic principles of language acquisition.

One of the most remarkable aspects of archaeological documentation relative to modern human beings in Europe is the presence of musical instruments. The oldest known musical instruments are the flutes of bone made 36,000 years ago in Geissenklosterle, a cave settlement located in southern Germany (D'Errico et al., 2003). Even though reduced to fragments, these flutes are quite different from those made presumably by the Neanderthal from Divje Babe. The Geissenklosterle flutes were made of the wing bones of a huge bird identified as a swan, and all have at least three holes for the fingers which are positioned in such a way that the space between the second and third hole is larger than the space between the first and second Straight lines, probably identifying the owner, were scratched on the surface using stone tools. It is likely that modern humans of the ice age who made these flutes produced other instruments as well. It is known for certain that they used natural formations as instruments: in several caves, painted stalagmites were found with clear markings that are the result of having been beaten (Macdougall et al., 2005). With technological progress over time, it was possible to make new types of instruments. All were made in order to extend the rhythmic potential of the body and the musical potential of the voice through the cultural material available.

The ethnomusicologist Kurt Sachs maintained that instrumental music began by using the body as a percussive instrument: slapping buttocks, stomach, thighs; clapping hands, and stamping feet (Sachs, 1982: 110). Over a long evolutionary period, humans substituted the beating of the body for objects found in the surrounding environment or made by hand. The beating of the hands was replaced by the beating of a wooden boomerang; rattles, made of nuts or animal hooves were attached to the ankles; beating sticks accompanied stamping feet. With exceptional originality and sensitivity, André Schaeffner[45] proposed a kinetic interpretation of the origin of musical instruments, partly suggested by the notion of body technique (Marcel Mauss, a contemporary of Schaeffner was working on body technique at the time).

Ethnomusicology and several brilliant studies in cultural anthropology —for example De Martino's (1961; 1980; 2000; 2002) – have shown that rhythmic time was a combinatory operator of the practice and complex uses at the basis of a community or a given society's experience and memory.

Kurt Sachs (1980: 21) defined dance as a 'rhythmic movement in a space-time succession'. He provided choreographic documentary proof – from the stone-age to the civilisations of the orient, from ancient Europe to the twentieth century – that this succession sequence models memory itineraries of the bodies of men and women. Dance practices incorporate the elementary threads that make up the spatial horizon of learning through important passages of human existence: birth, marriage, death, the sowing season, hunting, love, war and sickness. This is the space which separates the body and memory, between invention and the response of the cultural collective. Bergson (1983) maintained that 'time is invention or

[45] Schaeffner, A., Origine des instruments de musique. Introduction ethnologique à l'histoire de la musique instrumentale, Paris, Payot, 1936.

nothing', referring to the dialogue between sensations, bodies and emotions. This process lives and develops throughout the history of modern culture, that arsenal of psychotherapy, a matrix bathed in the transient time of subjective experience. It is now important to understand how the process described by Bergson evolves, how relationships are created – between emotions and emotions, bodies and bodies.

From time to rhythm – it is rhythm that constitutes the crucial problem of the new millennium. While with animals, as we have seen, there is usually a synchronism between periods of activity and periods of maximum physiological functioning, with humans such a synchronism might be reduced or even lost due to the complexity of the social structures in which they live. This de-synchronisation of social and corporal rhythms can cause the appearance of emotional disorders - somatic and interactive disorders – which are consequences of temporal stress to which biorhythms are exposed. Apart from being a conquest of consciousness, synchronism has become a social necessity. One of the new frontiers of thought on the network focuses on the synchronisation between social actors using electronic technology. The synchronism of a team in a regatta, the reconstruction of an autobiographical memory of a family event, the participation in a football match, demonstrate that synchronism strengthens interpersonal bonding and reinforces social integration. The domination of synchronism relative to time and place explains why dance is the chronospatial map that gives procedural (or implicit) processes access to conscience, through the contextual execution of its choreography. The art of dancing is different from other activities based on corporal movement, and yet dance movements can be found in other types of kinaesthetic movement. Sport, theatre, conversation, work that follows a determined rhythmic scheme, and athletics and dance practice, are fundamental categories that articulate behavioural movement whose modalities differ in temporal organisation associated with movement. Boiles (1978: 373) distinguishes rhythmic time from theatrical time[46] according to the following criteria:

If movement is subject to the exigencies of narrative time, movement is theatre, not dance (author's italics). Regarding this point, an interesting

[46] For example, the presence of a text in the Hawaiian *hula* or in the Indian *barath natyam* is only of secondary importance with respect to the rhythmic organisation, the time dynamics of the narrative do not interfere with the execution of the dance. Whether it is a question of movement or text, they are organised in rhythmic time and both are forms of dance. Similarly, a greater part of the text and movements in Japanese *kabuki* theatre are organised in a continuum of rhythmic time, and most *kabuki* is dance and not theatre.

example is mime. If movement is linked to a form of linguistic communication, for example the action in silent movies, kinesics is regulated by narrative time and belongs to theatre. If, on the other hand, the movements of mime are coordinated with some rhythmic signal, the kinetic movements become dance movements, not theatre.

Going back as far as the Greek *polis*, Boiles' invaluable analysis assists us in understanding why narrative time constitutes the frontier between dance and theatre.

It was George Colli (1969), historian of philosophy, who maintained that the birth of Greek philosophy occurred when confusion had arisen in the art of memory (Plato had begun to write in order to save Greek thought from being lost). Confirming Colli's convictions, the praiseworthy historical research of Yates (1992), di Rossi (1983; 1991) and Carruthers (2003) established that the social practices of memory were built during the history of Western thought (including 'theatres of memory').These were based on systems of images: from Greek and Latin oratory to Raimondo Lull's bizarre castles, from Giulio Camillo's extraordinary Theatre of Memory to Giordano Bruno's mysterious pilgrimages to London and Shakespearian theatre. In the theatre of memory, the 'body in action' is crystallised in the 'scenic presence' that fuels 'chimerical objects' of modern memory, from linguistic signs to the Internet era: survival of personified simulations, epiphanies or monuments to a faraway memory of performance, of figurative and verbal icons, of efficacious gestures and action sequences – some of which contain cultural value and profound significance.

Musical time with its rhythms becomes a paradigm, a metaphor – simply an 'analogical element with respect to the omnipresent rhythmic manifestations in the life of individuals and society (Gasperini, 2001: 119): those biological manifestations in the heartbeat, in respiratory movement, in perceptive and motor strategies, in well-consolidated routines, and in already-mastered skills and habits acquired with experience, make up the implicit (or procedural) fabric of the memory of each living being. It is worth noting that it is the 'body in action' in the learning context, it is the choreography that illustrates each difference. 'Why dance?' is a woven cloth, it demands total participation in every relationship, and the courage of one's actions to share when confronted with differences.

PART II:

BODIES AT THE MIRROR OF SOCIAL NETWORK

CHAPTER THREE

THE "SECRET HISTORY" OF THE BODY IN THE SOCIAL SCIENCES

3.1 The "Metaphor of Origins"

The body remembers and its memory is connected to our animal origins. The human animal is a social animal that by its nature needs physical and social relationships with its own kind to live and prosper. A life of interaction, common to man and animals, allows living beings to act upon real things by means of motor behaviours and to organize the outside world to their advantage.

As human beings, beyond perceiving the external and objective nature of the behaviour of others, we also experience ourselves as conscious and willing agents of events. The body is in fact the structure and the mode of our being, feeling, understanding and acting. Outside the body there is given no awareness and no world. Understanding is the body in action, and action consists as the spatial-temporal dynamism of the body. Husserl more than once shed light on the role that the body in action has on perceptive processes. To use contemporary terminology, we could say that, according to Husserl, there can be no perception of the other without perception of one's own body.

In modern culture, the dramaturgy of the body in action forms the basis of the theatrical presuppositions of the human being's capacity for performance. From 5th century Attic tragedy to the contemporary mobile interface, the stage actor activates, in the mask he wears, the "motor programme" that makes it possible to effect a given action with a given intention. The mask embodies the space of shared action. There was a time, many millennia ago, when theatre was sacred, ritualistic, a place of collective celebration with a shaman in the lead role. Without a doubt, his role was to bring the members of the group together; he made a fabric of the community, repeating chants, words, invocations and liturgies that narrated the heroic stories of a people, their people. Later, it is said that this conscious ability to create collectively a shared space was lost. Better expressed, it was shattered in "the time teeming with gaps and discontinuities" (Bachelard, 2009:19) that constitutes the fundamental characteristic of modern temporality.

There was a time, in the early 20th century, when a number of people were involved in creating theatre in different and faraway places. Some names such as Kostantin Staniwslaski, Jacques Copeau, Vsevolod E. Mejerchol'd, Adolphe Appia, Edward G. Craig come to mind. Something united them. They rebelled against the degradation of theatre at that time and reclaimed its ethical and artistic value. Mejerchol'd defined theatre as the "art of man", by which he meant that theatre is made by man, like the other arts' of poetry, painting, etc., in the sense that it uses man's body asinstrument and medium. As demonstrated by recent interdisciplinary research in neuroscience and theatre[1], the study of the body in action mainly concerns the work of the stage actor[2].In connection with this, the famous English playwright and director Peter Brook observed:

> The actors' work would be in vain if they could not share, beyond every linguistic and cultural barrier, the sounds and the movements of their bodies with the spectators[3], making them part of an event which they themselves have to help create.

The actor's body is the place in which it is possible to ask questions about the world "out of joint", unhinged, using language to narrate modernity, offering the performance of its laceration in the same language as the laceration itself. The actor's body is in fact the place of identity for the playwright and the engine for the staging. Whether it be Sophocles or

---

[1] Cf. Sofia G., ed. by (2009,) Dialogues between theatre and neuroscience, Rome, Alegre.

[2] Deriving from the Greek *théatron* (from *théaomai,* meaning "to see"). Theatre comprises the arts in which a story (a drama in the etymological sense of the word) is represented in the form of recited text or scenic dramatization. A theatre performance is carried out for a public, using variable combinations of words, movements, music, dance, vocalizations, sound and potentially every other element that comes from the other performing arts. Strictly speaking, the existence of a script is not a necessity: *the body's movement in a space, with the aim of narration and illustration*, performed for a single "spectator" is by definition "theatre".

[3] The spectator, also collectively defined as the "public", is the person present at an event, usually without any direct involvement. In theatre and especially in the performing arts, the artistic event is aimed at the spectator, but fruition is sometimes accompanied by more or less active participation in the events represented. Especially in theatre, spectators play a fundamental role not only as receivers of the representation, but also for the privileged communication established between themselves and the actor on stage.

Shakespeare, Camus or Goldoni, the actor's body is a container in which historic awareness and experiential vision come together, a space to be shaped and reinvented by continually drawing on an enduring knowledge of the pathways of art: a dreamlike but rational territory, it is shaped by the "once upon a time" past, staging ever-changing scripts. The actor's body carefully scrutinises photos of the past much as a lost explorer might scrutinise extinct maps of a country he does not recognise. Theatre is the bond between a human being – performer and human being – and a spectator, and in their continuous reciprocal feedback, they become one. Theatre embodies the possibility to set tremendous forces in motion, producing a psycho-physiological dilatation in the spectator through the performer's psycho-physiological dilatation. This is the "base level" of theatre.

In the past century, research was carried out to uncover the "actor's secret". Following the research initiated by Stanivslaski in the first years of 20th century, arriving at the end of the century with Grotowski, and continuing up to the present with Eugenio Barba and theatrical anthropology, attempts have been made to explain this base-level. The actor is on stage, attracting and holding our attention. We follow him without ever taking our eyes off him and we examine every little variation to understand its meaning. As spectators, we can only "react" from our seats: our backs, our adducting muscles, everything in us is in total tension, great or small, as if preparing to react but not knowing where to go or why. And still we remain immobile. As the actor voluntarily puts his own body, as well as his own primary emotions, into a precarious equilibrium, he makes himself the lead role and singer of his own rhythmic body, "the mode of understanding which, before any conceptual or linguistic mediation, gives shape to our experience of others".

The neurobiological approach to theatrical science is unable to "stitch" together the supposed gap between actor and spectator gap which supports the modern "language of signs" and which strongly influences the history of the body in western social science. Examples of this can be found in anthropologist Victor Turner's, (1986), posthumous work *The Anthropology of Performance,*[4] in John J. Schranz's research in collaboration with cognitive scientists Richard Muscat and Glymm Goodall, published as (xHCA (*Questioning Human Creativity as Acting*)[5]; also in Jean Marie

[4] The introduction to this work comes from Schechner's "Body, Brain and Culture"

[5] On the methodological positions and perspectives, see: R. Muscat, J.J. Shranz, *Performer as Act of Faith*, GHE, Malta, 2004; id. *Alla ricerca dell'uomo nonprogettato*, in "Teatro e storia", XVIII, n.25, 2004.

Pradier's work, which is defining a precise area of multi-disciplinary research based on a biological theory called "*Ethno-Scénologie*"[6]. No doubt: language – understood not so much as mere instrument of communication but as a deep cognitive function—is the thread of the spider's web we have woven around ourselves in an ecological and symbolic niche to which we now have to adapt, much as a cocoon with no glimpse of the butterfly. However, the *coup de téatre* is our rhythmic body: the body where the evolutionary *bricolage* is enacted "from pre-existing structures". The body is bearer of the constraints and compromises imposed by natural selection.But natural selection has allowed important natural innovations, and the continual, cumulative rewiring of the composite and imperfect engineering of our brains.And so it has let loose the unique possibilities of the intrinsically linguistic human cognitive universe, so that a large African biped mammal can now read, write and pilot a spaceship (Pievani, T. 2010:10).

And this rhythmic body, so fundamental to our equilibrium in time, space and the continually changing encounter with the world that surrounds us—where is it? And, last but not least, does it constitute one of the essential preconditions for establishing effective intercultural dialogue?

Looking at the flashy social shop windows of contemporary web sites and the panorama of editorial catalogues, the answer is certainly not thrilling. Nowadays, the body in action is depicted as the victory of its absence, wounded or dead. In its place reigns the so-called "body-icon", which the cosmetic, theatre and science industries celebrate as the unique, impenetrable secret area we all want to possess and where we all want to become the sole actor and interpreter. This body, lacking history and memory, has been invented on European and American soil in the 21st century, against necessity, against suffering, against time. And against the world, the world of nature, which is that of fate. This invented body has installed itself over our choices, both individual and collective. It has seized power (Juvin, 2005), becoming "normal". "Body icons" have been "promoted to high rank", since from the outset they are "malleable and potent" (Beegan, 2008), and therefore long-lived, adaptable and renewable in the grinding advance of time.Each different "body icon" is a bodily representation, explicit and tangible, circulated within a micro-formation of a given cultural universe. As are, for instance, the icons of leaders

6 Cf. Jean Mari Pradier, Théatre et Neurosciences. Ou l'interminable aporie in Teatro e neuroscienze (2007), ed. by F. Bortoletti, n.16:73-87; cf. also La Scène et la fabrique des corps, in id. Towards a Biological Theory of the Body in Performance, in "New Theatre Quarterly", Cambridge University Press, VI, 21, 1990:86-98.

exposed by the mass media to the global public, or the bodies invented by videogames. The "body icon" is the imaginary space[7], that furnishes the so-called "network cities", the "re-tribalised" communities, the "non-spaces" and the "hyperspaces", on which the social scientist's lexical fantasy has often focused its search for sense in our present day. It is thus the paradoxical place in which dreams become reality and projects are carried out, becoming enterprises that give physical reality to the ideal. What is more, the body is also the place where these ideals are transformed, enabling them to go beyond the limits of time, space and culture: nothing is impossible for this superhuman and ethically "pure" body. Indeed, its value as paradigm lies in overturning the hierarchy between truth and appearance; its associative function is the multiplication of the mirror of Narcissus. As pathway for appearance between the inner self, the outer self and its representations, the mirror of Narcissus constitutes the body's "beyond"—the victory of its absence.

The heroic, albeit illusory, search for unity that characterises Nietzsche's work laid the foundations for the modern body as "metaphor of origins"[8]. The most well-known researcher of the body in the social

---

[7] The term "imaginary" refers to a combination of visible and verbal images that are generated by a society (or part of a society) in relation to itself or other human groups. Theorists of the imaginary view the cultural apparatus as a semiological construction that gives its members the illusion of comprehending reality. Social dreams, expressed literally or figuratively, tend to substitute what we know with what is absent, with what we desire. The social history of the imaginary attempts to clarify the relationship between the object and the subjective and the objective world, the inner and the outer, the material and the psychic, always within the sphere of a community, even when analysing individual cases.

[8] In the history of western thought, the body is the privileged metaphor for conceiving the world, including its cybernetic and digital expression. Until a few decades ago, historians of ideas, philosophers, art historians and most anthropologists viewed this metaphor as more than a literary trope. It is a "system of images", relative to the historic era it belongs to. The primary function of this metaphor is to crystallise the bond between the body and objects in the representation it incorporates. The body and its functioning provide the metaphor that portrays and legitimises Rome's political organisation, as can be seen in the psychagogic effect that crowns the whole story.

The entire tradition of western political thought bears indirect witness to the effectiveness of the corporeal metaphor and of its normative function, starting with Plato, who constructed his entire *Republic* around this metaphor. Hobbes' state, the very emblem of modernity, is also a bodily metaphor: we can see this on the title page of *Leviathan*, where the state is a big, immortal sovereign body made up of innumerable small, mortal bodies of subjects. We cannot consider sovereigns, powers and bodies without making one further reference: in his famous book, *The*

sciences, Bryan S. Turner (1991; 208), also used the phrase "metaphor of origins" to signify the fundamental dimension the body takes on for human existence in defining its relationship with the world. Turner focuses on the originating character with which the body, here and now, exposes us and itself to the given world, establishing the very possibility for the subject to produce knowledge. According to Turner, the human body is the easiest "instrument" to use to transmit sense, which as such plays an essential role in the expression of the society as a whole. He sees the body as having an immediate capacity to express shared values, human sexuality and social power. Inasmuch as the body expresses sacred and sexual power, this control incorporates its sovereignty over sexual and religious expression. Our "incarnation" in the world is characterised by its precariousness, in which religions are the cultural ways through which we try to learn of our vulnerability by means of mythology, and by bodily practices that try to camouflage death, for example, by placing a mask over the face or by mummifying the body. Turner says that we can speak of "somatic societies" (1992) when the main political concerns and social unrest are translated into changes and distortions of the body image.

The body also forms an important bond between the individual and the state. Thus, in modern times the body has become a place of political conflict. The unbridgeable gap, regarding the possibility of educating the body in its most authentic sense, is determined by the idea of body conceived as "inert matter":a "body-object" that has lost its original sense of openness, of "co-exposure" to the world.Thus, according to Turner, the body is, in short, a highly conflictual expression of the emotional malaise that afflicts modern and post-modern man[9].

The "metaphor of origins", in which Turner sees the fundamental dimension of the body, forms the crucial link between nature and culture, individual and society, individual and power, "the worldly body of living" *(Leib)* and the "somatic or anatomical body" *(Korper)*, which makes the

---

King's Two Bodies, Ernst Kantorowicz (1957) uses a careful, detailed phenomenological analysis of the division of the king's body into two parts -- the mortal natural body and the immortal political body – according to the axiom of political theory that originated in the late Middle Ages and lasted through to the 20th century.

[9] Though cognisant of the ample debate - which cannot be reduced to a mere question of terminology - between those that oppose the supporters of "post-modernism" and those that support "advanced modernism" or "late modernism", we have chosen to use here the term "post-modern" to refer to the historical period that goes from the end of the 20th century to the beginning of the 21st century, in order to indicate its novel, ground-breaking elements.

body the "absent presence" (Shilling 1993) in both classical and contemporary sociology. In the 1980s there began to be a systematic interest in the body and the identifiable (even if inconsistent) *corpus* of knowledge, known under the label of "sociology of the body". This label has encouraged the formulation of detailed lists of the emerging phenomena that led to its use. These phenomena include the second wave of feminism, the gay movement, population ageing, AIDS, pornography, doping, new biomedical techniques (especially assisted reproduction techniques), reality and virtual communities, military and industrial cyborgs, and the culture of narcissism (Borgna, 2005: 52-102).

The volume of writing about the body has exploded in recent years in an attempt to match the literature produced by post-modernism, post-structuralism and post-colonialism on the world's contemporary cultural "landscape".As "actor" in so vast a discussion, the theme of "body" has played a central role in the analysis of gender, identity, race, science and technology. However, the literature that has emerged is of a theoretical nature, having almost completely ignored empirical and experiential analysis. The identified types include models of "discursive" and "material bodies" (Turner, 1984); "physical", "communicative", and "medicalised" bodies,"consumer", "individual" and "social" bodies (O'Neill, 1985; Turner, 1991). Even if this proliferation of possible roles and functions might seem to indicate an increased interest in the body, Frank (1990:160) rightly states that these categories are by and large functional to academic agendas rather than being of real use for understanding the body's materiality. He writes that "in all this research, the body itself slips away, lost in the social practices that incarnate our predisposition and understanding".

In the "post-traditional cosmopolitic" world, to use Antony Giddens' expression (1999:103), the "secret story of the body" (Turner 1991:12) models so-called "somatic societies" (Turner 1991:12) as "experiments in living" (Stuart Mill, 1985), founded on the metaphysics of the "subject". There was a change in the 1980s that began the phase that we are still living in today. As affirmed by George Steiner, theorists and artists seem to dance hilariously and happily around a meaningless ark. This is the attack launched against the "subject" and his cognitive responsibility. According to Foucault, the "subject" is an empty space that fills up according to the impact of devices that act upon him from outside, until he himself becomes a device. Deleuze goes further, transforming the subject into a "body without organs that grows rhizomatically" and therefore is a neutral "subject", with no responsibility, either towards truth in the Platonic sense or towards his own desire for knowledge. He is, in the

Italian version given by Gianni Vattimo, "weak". The outstanding champion of this "subject" is Nancy's (2007:23) so-called "*Corpus Ego*", incarnated by the contemporary "flying bodies" picked up in the streets, the megalopolis, the outskirts, in transitory places and places of surveillance, in commerce and oblivion. According to Nancy:

> The body is neither significant nor meaningful: it is exponent/exposed: *ausgedehnt*, the extension of the break-in which is existence. The extension of the *there*, the place of break-in from which one can *come into the world*. Mobile extension, spacings, cosmopolitan and geological gaps, drifting, sutures and fractures of the continent-arcs of sense and the immemorial tectonic layers that move beneath our feet, beneath our history.

In these bodies, we witness an aesthetic over-determination. As pictured by Nancy, the body is a "place of writing", the skin a surface on which a person can write his story. Tattoos are exhibited to onlookers as incarnated books. That is to say, the body and the skin are used as personal works of art. "For me giving my skin to a painter is giving myself to art", in the words of a Roman musician with an arm full of many-coloured tattoos of women's faces, which replicate the work of a famous Mexican artist of the Twenties. All data indicate that tattooing is continually on the rise. In western society, tattoos are used by important media personalities, like actresses and footballers.They have apparently lost the original function they had in traditional cultures[10]. Above all, tattoos represent the *aestheticization of the body*, one of various aspects of consumer culture. The imitation of motifs used in other social and cultural contexts (like Japanese or Maori tattoos) is only apparently linked to those traditions. Body markings are perhaps amusing or ironic, but they generally no longer communicate a particular meaning. But this is not always the case. For

[10] A "tattoo" is a permanent artificial deformation of the skin found in many traditional societies, which used them as initiations as well as for purely aesthetic purposes. This deformation, made by pricking, is largely used throughout Europe by lower social classes, and in the Near East. In Italy, the tattoo has various regional names, such as "mark", "*marconzito*", "sign" and "devotion". There are two main types of tattoos. The first are scars, mainly found in dark-skinned people (e.g., in Australia and Central Africa) but also found among the Pigmies and the Andamanese. These tattoos consist of deep skin incisions in the body and the face, after which certain substances are inserted in the wound to delay the scarring process, thus increasing the volume of the scars. The second type of tattoo are pricks in which a bluish (or other colour) substance is introduced into the wound. These are used by light-skinned people, especially in Polynesia, Northern Africa and India.

adolescents, for example, “tribal” tattoos and *piercings*, recall a sense of belonging to the “tribe” (of friends, or music fans, or simply of people of the same age).

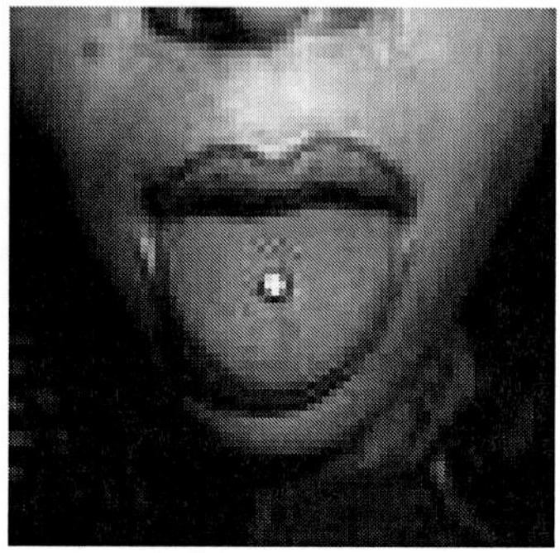

Figure 3-1: Example of a piercing

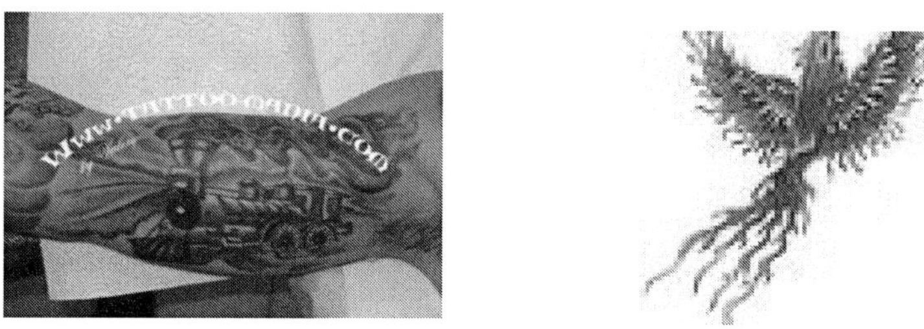

Figure 3-2: Samples of tattoos

The metropolitan tattoo is a signal of recognition that expresses a common lack of contentment with one’s own “natural” body, as well as the willingness to do something about it, to modify it culturally, even if this entails suffering. This body becomes a hybrid: objects and products are introduced in it that give it a compound identity and are part of the representation of an improved body image, personally chosen and individualised. The “subject” acts on his or her “form”. The body can be deformed or informed. It may be disfigured/mutilated or bestowed with a meaning. Deformation gives the body a social mark, whose appearance is controlled through diet, clothing and piercings. Information is a technique that marks the body’s surface with an identifying sign whose purpose is to make oneself known and recognised. Equipping the body with new and different technologies, the “subject” brings about changes to this hybrid body with the illusion of renewing its appearance. Hybridisation alters sensations of the self and the world, as well as motor capacities and mental processes. A curious representation of this "implanted body" was given by David Cronenberg in his film *eXistenZ*, where a videogame directly connected to the player's central nervous system transports him body and soul into virtual space. And it proposes a continual becoming, since the operation can be repeated according to the player's desires and to the possibilities offered by biotechnological progress. In hybrid identity, matter makes body with the body. The mutant body seeks its own rules. It is no longer only a matter of designing one's own bodily appearance for oneself, but also of achieving personalisation through tattooing, piercing, implants, prosthetics and food supplements. The hybrid identity differentiates itself from others with its extraordinary artificial body. This presumed "creativity" achieves new symbolic structures. Hybridisation produces mixed structures with techniques that denature functions, objects and relational modes. Hybridising oneself shows that the body can be shaped: biology can become a "creative" interface, going beyond the rules.

The above-mentioned social practices indicate a use of the living body as raw material to be moulded in the iconic artifice into an elaborated being, reproduced and spread by multimedia networks, from film to digital techniques. The point of reference is no longer a shared cultural model. What is at stake is no longer only staying young and beautiful. Rather, it is turning the body into an empty page on which we are free to draw the mental image we have of ourselves. This "liberty" is, however, not without risks. As the sociologist Patrick Baudry (1991) explains in his book *Les Corps Extrêmes*, man's main hazard is his own body, against which he has to fight in order to overcome his weaknesses. And indeed, there are many people that fight against nature with every possible means, undergoing daily beauty treatments, dedicating themselves fanatically to sports, or following extreme diets. It is as if we found ourselves in the presence of a two-way movement: the destruction of the body on the one hand and the attempt to reconstruct it on the other.

Body image[11] has the function of a playful staging of the self and of eroticisation of the body. According to Elizabeth Grosz (1994:83), body

---

[11] The image that we form of our body is based on complex, conceptual representations. And it can contradict our aspirations. The direct knowledge of our body state is made possible by the proprioceptor system. This includes numerous systems whose function is to represent a wide range of information, from balance to temperature to possible bodily harm. Sometimes this information is conscious (such as with pain), other times it is unconscious (as in information regarding balance). They make the body a special "object", which is at the same time a

image constructs the dynamic interrelation between the physical body and its social extensions, such as clothes, make-up and underwear: on this view, material or crafted "gender signs" are the means of connection between body, mind and context.

Attention to body as the interface between the individual and the social, the biological and the symbolic, illustrates the theoretical framework that characterised the genealogy of sociological reflection on the meaning and social use of the corporeity starting from the 1980s. According to these interpretations of the body, man's instinctive automatism, though it has not totally disappeared, has gradually diminished over the years, being almost entirely substituted by the impact of cultural models and orientations. This thesis has, however, been amply refuted by the findings of current experimental research in the fields of neuroscience and anthropology. No body means no interaction. It is not the body that speaks various languages like gestures, curtsies and eye expressions that is referred to here, even if these are all undoubtedly important codes for denoting meanings of communication which differ from one context to another. This "linguistic" body has been analysed extensively by the so-called "interpretative turn" that has been brought to light in the most illuminating fashion by Geertz (1988). Instead, we are referring to the body that composes the space of interaction between one human being and others, as well as between human being and non-humans such as animals, plants and things. As demonstrated by research in the area of non-verbal communication, variations in the distance between bodies govern the space of social interaction, organising the exchange like a "dance" (Birdwhistell 1970), whose purpose is to model bodily movement in the actions that are found the various "pictures of experience" (Goffman, 1991).

It is a question – as the physicist and physiologist Ernst Mach demonstrated over a century ago, followed some time later by Henri Poincaré (1913)—of the body with respect to which we *map thespace that surrounds us* by means of "prehensile movementlooks and locomotion" (Mach 1905:342).

This space, attentively treated by Marcel Mauss as the notion of *body technique,* is even to this day, as in Poincaré's time, "uneducated", though

---

source of information accessible only to the social actor, the centre of his action and the point of reference with respect to which his perceptions are organised. The Austrian psychiatrist Paul Schilder (1886-1940) had already analysed the plasticity of the body image, showing that this image can be modified, both by techniques like gymnastics, dance, meditation, piercing and mutilation, and by clothing, which also forms part of the body image.

sensory information is an integral part of our everyday life and our working memory. Thus, it is especially relevant to consider the body's rhythms as the *primary horizon* of action and social interaction.

3.2. The Body in Social Theory

A predisposition to being very social is a human characteristic. Many animals enjoy a certain degree of social organization, but none benefit from it to the same extent as do we humans. As the size of our brains increased, so too did the size of our social groups. Something made us curious about others and want to live and co-operate in groups. Richard Wragham has a fascinating theory about the role that cooking played in causing such a radical change to the life of primates. Other theories include the need to keep predators at bay, or the search for food. Still others maintain that our more sophisticated intellectual abilities developed so that we could adapt to our new social situation.

Understanding what it means to be a social animal is fundamental to fathoming the human condition. As we progress, we realise that our social nature is deeply rooted in our biology, and not only with regard to the cognitive theories that have we about ourselves. In fact, we begin to realise that it is our bodies' equipment that helps us find our way through the labyrinth of social relationships.

But where is the body? And what is its function within social theory and practice?Everyone, for either professional or personal reasons, is asking questions about the bodies of both men and women. Instead of continuing to speculate on the soul, passions, or the phenomenology of the spirit, philosophers now choose to concentrate on the reality of the body and the limitations of the human condition. Sociologists and anthropologists are convinced that the existence of man is primarily about the body. They know that every society designs its own particular knowledge of the body. Psychoanalysts now agree about the importance of the body in determining health and disease, and recognise that physical suffering has an unrelenting, even obsessive effect on both neuroses and psychoses. Doctors, who have to know how the biology of a body works, are also aware of the fact that the "*body/subject*" defiesmaterialistic reductionism. Jurists, who are sensitive to the acute perception that the physical dimension has reached in modernity, recognise the need to draw up a new legal statute for the human body.

In short, whether for work or social commitment, we are all forced to deal with situations that require a comprehensive and inter-disciplinary approach to the body. Last but not least, the body also concerns a curious general public, aware of the difficulties and unease created in our day by the paradoxes linked to media over-exposure (the press, advertising and pornography), and by the nullification of the bodily dimension in social networks (internet, videogames, blogs, etc.). It is not a question of modifying the entire history of the body and its conceptualisation, nor of re-establishing the paradigms that have determined our thinking about the body. Rather, our task is to put into focus current orientations in order to present the critical literature on the question.

In *The Body and Social Theory*, Chris Shilling (2003) wrote a comprehensive critique of recent sociology of the body, in which he showed that while research about social theory had put the body on its agenda, it had nevertheless failed to focus on the body as a specific subject of analysis. Shilling noted that the current state of the body in social theory is part of and a consequence of traditions that have failed to consider it as a "material component of social action" (Shilling 2003:70). In fact, though it sparks increasing interest for the communicative dimension of social relationships, the body undermines the prevailing model that, in summary, was first developed by Durkheim and later by Parsons in sociology and Lévi-Strauss in anthropology. According to this model, the various forms of life and social organization can be traced back to complex and in some ways invariable formulas. It has therefore become a cliché to see a tension within classic social theory, especially between Marx, Weber and Durkheim. Marxist-influenced sociology focuses on social relationships between groups and classes, whereas liberal sociology has concentrated its efforts on interactions among individuals, rather than on questions of "embodiment"[12]. That is to say, it concentrates on the ways in which material artefacts of the mind – or "social facts" – opinions, faiths and identification of socio-economic and cultural groups – can be carried out by individuals and acted on collectively. In this sense, it has been far easier to study the "differentiated" body and defend its rights and its "ownership" using the theories of liberal individualism than to use theories based on interactions among individuals and/or groups.

Until recently, both lines of thought largely took the body for granted, especially concerning the construction of action. As a result, there is widespread opinion that when it came to the body, classic sociologists had little or no interest. In almost all the articles or essays about the sociology

---

[12] For an exhaustive interdisciplinary explanation of *Embodiment*, see: *Perspectives on Embodiment: the Intersections of Nature and Culture*, ed. by Gail Weiss & Honi Fern Haber, (1999), Routledge, New York and London; and Kate Kregan, *The Sociology of the Body: Mapping the Abstraction of Embodiment* (2006), Sage, London Thousand Oaks-New Delhi, specifically parts II &t III

of the body, any reference to the body in past theories is dismissed as an "*absent presence*" (Shilling, 1993). After all, the term "body" hardly ever appears in the analytical indexes of fundamental sociological works, except for Pareto's *Trattato di sociologia generale* and Parsons' *Social System*. There is no reference to it in the indexes of the translations of works by Durkheim or Weber, nor does it appear in the analytical index of Marx's *Das Kapital*. Traditional, old-school thinking undoubtedly made an uncertain contribution to the analysis of corporeality; at the very least, it was not homogenous, with the exceptions of the brief essays by Simmel in his *Sociology of the Senses* and Mauss in *Techniques of the body*.It was not until contemporary times that the body became an object of reflection for the social sciences; specifically, with Turner, who dedicated an entire book to it in 1984, followed by Berthelot (1985), Featherstone (1991) and Shilling (1993).

The body was only a marginal presence in traditional sociology in as much as it was the negative pole of the Cartesian dualism that was the basis on which the discipline was originally developed[13]. The dichotomies of nature/culture, man/woman and public/private combined and positioned the body as the pole in opposition to the so-called "rational mind". Turner (1991) demonstrated that the classic theorists, such as Marx, Weber, Durkheim and Simmel, who had established the initial cannons of sociology, were only interested in understanding the public aspect of society so that they could analyse and criticise the systems of social order. In sociological terms, the result of the disjunctive logic that introduced the well-known distinction between *res cogitans* and *res extensa* from the ancient Greek model, created a paradigm that reinterprets this "disjunction" as a problem regarding social control. Within this interpretation of culture as a system of laws and coercive customs, corporeality becomes the "object" of selection and manipulation of the sphere of instincts and natural needs. Although Durkheim, Freud, Weber, Elias and Foucault all had different points of view and different theoretical premises, they all believed that this control constituted the fulcrum of sociality. For Durkheim, Freud and Elias, it is in fact the laws of morality of "external" civilisations that must regulate and moderate the virtually infinite flows of needs, pleasures and impulses that are part of the nature "within us all" (Stella, 1996: 20). What follows is the progressive exclusion of the body's "animality" from public areas of interaction, which, according to Elias

[13] Descartes recognised the substantiality of the body and the possibility of founding a study of the body without referring to spiritual elements. From this, it is only a small step to seeing it as a mechanism. Corporeal substance is characterized and determined by its extension in length, breadth and depth.

(1937a: 346), determines an increasing "privatization" of all bodily functions and a progressively clear-cut separation between the intimate and public areas of human life (Elias, 1937a; 1937b). The "disciplined body" describes a style of use of the body as well as a bond between the body and objects that is based, ideally, on four elements—control, desire, relationships with others, and relationship with the self.Thanks to control, the "disciplined body" becomes "predictable"; when faced with desire, it is "lacking"; in relationships with others, it "closes in on itself"; in relation to the self, it is "disassociated". Foucault's concept of a "socially constructed body" is a pertinent example of a "disciplined body". As Turner (1984: 245) notes here, according to Foucault the body denies the "immediacy of tangible personal experience, which is instead contained in the notion of 'my body'".

Turner's observation reveals the sharpness of the conflict between the individual and society which, in the body, inhibits the interactive construction that is the ongoing result of an effective exchange among people, their cultures and their histories (languages, myths and customs). This conflict incarnates the "disciplined body" and certainly as a concept it is a "convincing attempt to classify the modalities of the body as social construct" (Borgna, 2005: 48). By "body as social construct" he means:

> the result of interdependence and interweaving between the social construction of the body (processes, methods, structures) and contents through which a social interaction or formation acts in an organized and constant way on the morphology, physiology and behaviour of the body and/or its parts.

Foucault's work was aimed at analysing the way in which political technologies invest our bodies, our health, our food and housing, our living conditions and, in short, the entire space of our existence. His "disciplined bodies" constitute the development of history in modernity through the fields of medicine, pedagogy and demography.Foucault (1978: 127) writes:

> For millennia, man continued to be seen as he had been by Aristotle: a living animal who is also capable of political existence; modern man is an animal whose politics crucially involve his life as a living being.

Foucault (1976; 1986) shows that during the classical era of modernity, a new concept of life in society appeared, which profoundly changed the principles of how to manage the body.Power was no longer expressed by death sentences or the deprivation of life. Little by little, as the desire to put obstacles in the way of life disappeared, power tried to nurture life and administer it on the basis of new sources of knowledge. A political arithmetic tried to define and measure the strength of different nations according to birth-rates, conditions of health and other statistics (life expectancy, and the evolution of weight and height). Political power took responsibility for the lives of the population, because the health of bodies became part of the inventory of the wealth of competing states (1976: 183-184).

This point of view presupposes new principles of efficacy, ownership and identity that progressively free themselves from the burden of metaphysics and the juridical models of former regimes. Social control no longer rested on the appropriation of bodies, but on one's condition as citizen. As it turned into the subject of urgent interventions aimed at improving its efficiency, the body became a source of worry. From being tortured, flayed, beheaded, amputated and killed, with its suffering and death provided as public entertainment, the body became a recluse, hidden away like a prisoner. The transition from the torture of the body to imprisoning it signals a reversal of the public/secret relationship. Not only were gallows, stocks and wheels eliminated, but from the end of the 19^{th}century until 1981, the year that the death penalty was abolished in France, the carnage of decapitation was hidden behind prison walls.

In this evolution, there is a movement away from a secret order that resulted in public agony towards a public order associated with a punishment hidden away from public view. When recovery and rehabilitation of the convict is now the goal, the techniques are no longer aimed only at the expiation of error; it is not a question of trying to hurt the body, but rather of suspending its rights. The body is still in play, but as an intermediary tool with the inner self of the guilty person in order to correct his or her "soul". This new objective requires specific techniques of detention and observation. The specificity of power is no longer that of showing itself and overshadowing everything that is dominated. The regime of "visibility" has been overturned.Power no longer seeks the light, but is content to stay in the shadows. (Deleuze 1975; 1986). According to the "panopticon model" (Foucault, 1975: chapter III), the individual must feel permanently under surveillance. In the beginning, the Panopticon – an architectural project for a prison by Jeremy Bentham[14] – comprised a structure with a circular circumference and with a watchtower in the

[14] Around 1794, Bentham came up with the *Panopticon Plan,* a design that would make it possible to design prisons that could guard inmates effectively and economically. It had already existed for two years when he decided to use the design for his factory, in which prisoners were made to work, although in the end they had to be replaced with paupers. (See also *Gilbert's Act*).

middle (figure 25). The outer structure, which is divided into well-lit cells, is home to the prisoners, in full view of the guards who are inside the watchtower but out of sight. The system achieves discipline: separated from the other but in full view, the inmate feels constantly under control, whether he is or not.From the time he gets up until the time he goes to bed, he believes he is being watched, and this makes him submit to power, keep himself in check and in a certain sense be his own guard.

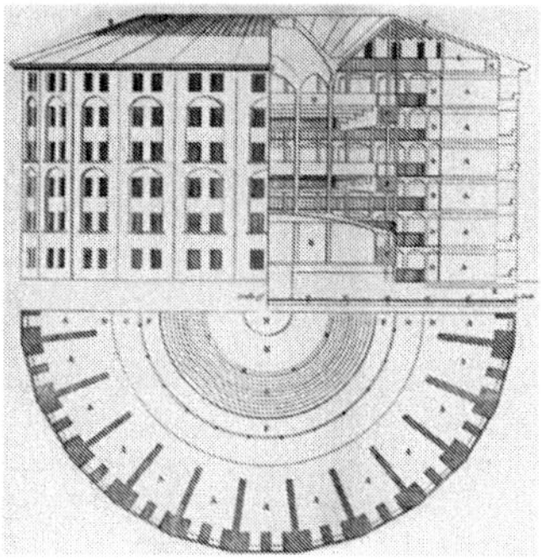

Figure 3-3: A Panopticon

According to Foucault, the *Panopticon* is the metaphor of the economy of modern power. In this kind of discipline, the law is substituted by the rule, coercion by internalization. This architecture corresponds to this kind of discipline: space is fenced in, enclosed and with boundaries. Everyone is closed in his or her place, and any attempt to make a move puts life at risk, given the possibility of contagion or punishment. Modern power is actually a *bio-power*, whose workings affect the body in two ways: global management of the population on the one hand, and the taking over of the most intimate details of an individual's daily life on the other. For Foucault, this division of power consists in one form of management based on regulations governing the entire bio-political arena, and in another form of management which is disciplinary and individualized in nature. His explanation of the power relationships that control the body has led to many approaches based on the political implications of his theory. Foucault also developed Sartre's work on "the eye", claiming that the possibility of being observed by an unknown and ever-present "eye" makes the actual presence of this power superfluous in order to discipline bodies. However, his considerations on resistance, as explained by the so-called "technologies of the Self"[15], keep the body-Self dualism, in that he sees the body as the manifestation or result of the position of Self within the thought process that is shaped by the microphysics of power.

In adopting the expression "the body", social theory meant "a language we are spoken out of", a "language of nature", in which we betray what is hidden because it is the least consciously controlled or controllable, and whose messages, both perceived and not perceived, contaminate and over-determine all deliberate expressions, starting with the word. This idea of the body as a "pre-discursive phenomenon" which nonetheless plays a key role in perception, cognition, action and nature, gave rise to the body's being considered a conflicted entity—whether as the organism studied by the natural sciences, or as the body perceived by phenomenology, or as the instinctive and repressed body of psychoanalysis.

The progressive substitution of the term "body" by the term "embodiment" reveals the new trend that is establishing itself in the international panorama of the social sciences: to consider the body as author and actor in the various ways of living in the world that link social order and the memory of a group or a given society on both a physical and symbolic level. Anthropology and ethnology have already had a significant influence in advancing this trend. The contribution of these disciplines to the study of the body is, unlike classical and modern sociology, both consistent and varied. "Can anyone think of the world without distinguishing between culture and nature, the 'tamed' from the 'wild?' Anthropologists and ethnologists have been trying to answer this fundamental question, which encapsulates the progress of modern science from Newton to Darwin, in order to organize pragmatically, in their field work, continuity and discontinuity between man and his surroundings. The interactive picture of traditional, as opposed to modern and post-modern societies, presents itself as both immediate and corporeal. Deemed to be inferior in the first series of *l'Année sociologique*, these societies were later described as being "archaic" or "undistinguished" in the same journal in the 1950s. When this absurdly a-historical viewpoint was abandoned, a

---

[15] Foucault's proposed that subjectivity is a process of craft-like construction (*τεχνή*: art, opera, work). This led him to redefine Socrates' imperative "know yourself" and to re-examine ancient philosophical practices based on "technologies that allow individuals to carry out, personally or with help from others, a certain number of operations on their own bodies and souls – from their thoughts, to their behaviours, to their way of being – thereby managing to transform themselves in such a way as to achieve a state characterized as happiness, purity, wisdom and perfection". (Foucault, M (1992) *Tecnologie del sé*, (Technologies of the Self) Turin, Boringhieri).

new anthropological trend developed, unafraid of studying both so-called archaic societies and modern ones. And even if the authors did not choose to study the body as their preferred subject of investigation, they did deal with it, describing it in ways that were more or less explicit. For example, Roger Bastide's most significant contribution (1995; 1996; 2001) was the study of collective memory as cultural *bricolage;* while Roger Caillois (2000) studied the relationship between games and man; and Morin[16], in his ambition to establish "genetic anthropology" as a discipline, attributed a significant function to the body in his investigation of the imaginary.

The first serious challenges to the nature /culture[17] debate appeared in the radical social and cultural climate of the 1990s. Mary Douglas (1970; 1973; 1975)[18], a British anthropologist, formulated an analysis that attempted to break away from the connotations of the body as "dirty" and "animal-like", proving that these types of associations were not based on real conditions, but rather on accepted social definitions imposed by the dominant way of thinking. Douglas suggests that these associations spring from the view of the body as "disorderly material", in order to propel it

---

[16] Cf. Morin E. (2001), *Il metodo 1. La natura della natura*, Raffaello Cortina, Publ. Milano Raffaello Cortina; (2004) Il *metodo 2. La vita della vita,* Milan, Raffaello Cortina; (2007) *Il metodo 3. La conoscenza della conoscenza*, Milano, Raffaello Cortina; (2008) *Il metodo 4: Le idee: habitat, vita, organizzazione, usi e costumi*, Milan, Raffaello Cortina; (2002) Il *metodo 5. L'identità umana*, Raffaello Cortina Publ.Milan, Raffaello Cortina; (2005) *Il metodo 6. Etica*, Milan, Raffaello Cortina; (2002) *Il cinema o l'uomo immaginario*, Milan, Feltrinelli.

[17] The concept of nature was finally defined in the 17th century. At the time, the idea of cultural collectivism did not exist. Man existed and as an individual, and he transformed nature. Rousseau also saw no contradistinction between nature and society. The notion of culture, as we understand it today and as we now believe to be universal, was actually formed in the 19th century in Germany. In his book "Power and Civilization (1983b), Norbert Elias explains that there was a very particular reaction in this region against the idea of universalism as expounded in the philosophy of the Enlightenment. For further in-depth analysis, cf. Bateson & Mead (1942); Hall (1969); Efron (1972) and Birdwhistell (1973).

[18] Mary Douglas studied under Evans-Pritchard and had an abstract vision of anthropological practice that was considered to be a re-elaboration and new perspective of the theories put forward by Emile Durkheim. She took a Ph.D. in *Social Anthropology* at the University College of London in 1965. The Royal Anthropological Institute viewed her as one of the most important figures in contemporary social theory, and she received many academic awards in England and abroad during her long life, including becoming a Dame of the British Empire in 2004. Another of Douglas's contributions to contemporary anthropology is undoubtedly the work she did on the concepts of risk and danger, based on a comparative analysis of religions and symbolic systems.

beyond the confines of culture. She challenged that idea with an alternative theory, namely that of "the two bodies" (one physical and the other social), as a means for defining body symbolism as a social vector of meaning in any culture.

The most relevant current contribution to the debate comes from Philippe Descola, a student of Lévi-Strauss and presently Professor of the Anthropology of Nature at the *Collège de France* and the Director of Studies at the *Ecoledes Hautes Etudes en Sciences Sociales* in Paris. This French anthropologist's contribution takes shape above all on the *methodological plane*.[19] Descola did his field work on the adaptive relationships of the Achuar Jivaro Indians of the Upper Amazon, whosesurroundings included forests, large predators and gardens with lush and varied flora and fauna.There Descola mapped the ties that enable the daily encounter between humans and non-humans, giving priority to a combinatory analysis of their contingent modes of relating. He interpreted the ethnographic present in accordance with contingent circumstances, intended as an instantaneous way of comprehending a given community in a given moment of its course.

Descola's field studies showed that for the Achuar Jivaro the techniques used to exploit nature have a clear link to social practices. The diversity of the classifying indices they use to understand the relationship among organisms reveals a flexibility of frontiers in the taxonomy of life forms. The Achuar certainly establish distinctions among the entities that inhabit the world. The resulting hierarchy of animate and inanimate objects is not based, however, on degrees of distinction among beings, nor on diverse appearances, nor on the gradual accumulation of intrinsic properties. Instead, it is based on a variation in the networks of relationships that is authorized by the acquisition of learning about perceptible qualities, which are unevenly distributed. Inhabitingthe jungle, the rivers and the gardens of their "animal brethren", the Achuar Jivaro

[19] At the time he was working on his ethnographic research in the Upper Amazon, two significant paradigms that dealt with the relationships between a given society and its natural surroundings already existed and were particularly alive in Latin America. One of these was a sort of geographical determinism, fairly concise, known as "ecological materialism" (or "cultural materialism") developed in the USA, primarily thanks to the work of Julian H. Steward (1902-1972), and based on the conviction that all the peculiarities of a given society could be explained by what characterized its adaptation to environmental constraints. The other current was a distorted interpretation of Lévi Strauss's structural anthropology, which tended to treat the environment as a lexis of properties that were rearranged in metric systems, classification systems and in ideological productions.

have "tamed" an otherwise "wild" world in order to make their own habitat comfortable and to establish an on-going equilibrium between their bodies and their surroundings. To achieve this, they have rejected the techniques of accumulation that, by contrast, structure the hierarchical societies living in a string of densely populated villages in the flood-plains of the Amazon (1986). Indeed, the symbolic framework that provides the rationale for their use of nature was not capable of absorbing the reorientation of social relationships that this choice would have brought about.

More than twenty years of field research among a small section of a society in the Upper Amazon, overlooking the turbulent waters of a muddy river, surrounded by tens of thousands of hectares of trees, moss and ferns, proves that every type of presence in the world, any kind of bond and any way of using this bond, constitutes a specific pact between the data of sensory experience, which is accessible to everyone, and a mode of aggregation of existing creatures that is adapted to historical circumstances. As Descola affirms, the frontiers of humanity do not stop at the gates of the human species. Therefore, analysis of the interactions among the inhabitants of the earth can no longer be earmarked to the sector of the institutions underpinning the lives of men, as if everything that is declared "exterior" to these institutions were nothing more than an anonymous accumulation of "objects" waiting for meaning and the determination of their use. Descola's contribution substantiates the tie between emotional intelligence, behavioural plasticity, speed of learning new things, and flexibility of the schemes of interaction that connect animal and human sociability. This can be encapsulated in the following question: "*Why is that given social fact, that belief, that use there and not here?*" (2005: 533).

This question provides an invaluable map for continuing theitinerary of our research. The salient answers to this question are written in the ways of living in the world composed by the body within its context of action. This is how it progressively accumulates, though in very different time-scales, the "hierarchy of wisdoms" (Simon, 1996) that shapes the relationship between nature and technology: the wisdom of the Phylum[20], the wisdom of the species[21], the wisdom of the group[22] and the wisdom of the individual.

---

[20] This form of knowledge is expressed as a set of brain processes (genetically coded and transmitted) that are activated by certain stimuli and situations. This type of wisdom is the result of millions of years of experience in the evolution of mammals, and is expressed in humans as the basic emotional response to certain stimuli and also as basic perceptive discrimination.

[21] The wisdom of the species is a culturally coded and transmitted combination of

3.3 An Apotheosic Body: the "Body Icon"

Not just wrapping. Also sophisticated *packaging* to promote ourselves. The body is the mystique of contemporary society. We want to be beautiful, healthy and immortal. The "body is everything": it can be found everywhere. Nude, photographed and exposed. It is a cult that has to be managed and manipulated: using scalpels to correct it, sports to improve it, medicine to cure it and to freeze it in time.

Figure 3.4: Volunteers pose naked for the artist Spencer Tunick

The soul's double can't be blemished. It is a lexicon that is more complicated than familiar—we use it without knowing it very well. Contraception, breast feeding, obesity, handicap, piercing, a harem, a webcam, the press and shamanism.Francis Bacon, Fellini, Barbie, and Pinocchio. In a society that likes to mix and confuse languages, the body has many words to narrate itself. It is a central theme of our times, but most of us do not understand its meaning. Memory and seduction are interwoven to provide representations of politics, art, power, faith and transgression. Reference to the body engages us in discourse about its rights (*habeas corpus*), differences (gender, ethnic belongingness) and forms of stigma (poverty, exclusion and disability). The body, object of

---

categories that allow us to analyse the world in an adaptive manner for the species. This type of wisdom has acquired thousands of years of human experience, and is expressed through language and other symbolic systems available to us.

[22] The wisdom of the group is the collection of skills and competences possessed by a group of individuals with a shared background (such as members of the same profession), which allow us to carry out complex tasks that most people would find daunting in a relatively natural way.

desire, expression of charisma and privileged space of public aesthetics, has also become a show. It has even been transformed into a weapon (the *kamikaze*).

Body icons have given life to the powerful social narratives of late modernism. They have contributed to the construction of the symbolic apparatus that the media has elaborated, reproduced and disseminated[23]. In this way, the body has become a supermarket, a cathedral and a monument to our present day. Inside, practices and meanings are mixed, along with commerce and symbols. Everything that we watch, read, consume and think has to do with the body, or rather, has to do with our will to control it. Showing off a body that is more than pleasing is the strongest demonstration of our capacity to manage life. A fit body guarantees human and professional success. Protecting it from the signs of time and keeping it active is a commitment and an important mission, the very aim of our existence. In this whirlwind of images, what escapes society is "the living man, man in flesh and bone" (Febvre, 1962: 544), that is to say *dominion over the senses, places and times* that shape his being, *a body among bodies*.

In the age of "the tyranny of intimacy" (Sennett, 2000:112), the body is no longer viewed as an immediate given or natural fact. It is referred to as the mindful body, the body as capital to invest in. It is also referred to as a "mental body" whose meaning is determined by individual or collective projects. This is the confusing mixture between fact and show that forms the *referential performance* of the "body icon".

The "body icon" is a construct that is exclusively limited to western written culture, aiming to govern the relationship between man and time. In it,the relationship with time is transformed, and the way we construct experience and biography, original knowledge and social rhythms,

---

[23] In this respect, Charles S. Pierce identified three types of signs, namely the index, the icon and the symbol. The icon is the sign in which some of the object's characteristics are represented, such as a mental image, a painting, a diagram or a metaphor. According to Eco, it resembles the object. To give a fitting example, icons are images of sports champions that have demonstrated their outstanding technical abilities with their malleable bodies. In contrast, the index gives evidence of the existence of the object it implies, without describing it. A cigarette butt in an ash tray reveals that somebody has been smoking in that particular environment, however, nothing is known about the smoker or the specific circumstances. The symbol is a conventional sign that stands for something else according to a corresponding code; it is a sort of 'law' (even if the word was banished from the sociological lexicon during Weber's time) that indicates the object with respect to a norm, but nothing is known about the object's existence or qualities. In a photograph, for example, all three signs can be present.

changes. Within the framework of the "space of flows" (Castells, 2002:460), the "body icon" composes time as the "absolute now" (De Kerkhove, 2003:26), that "shrinks the map of the world" (Harvey, 1993: 296) and ultimately transforms it into a "living skin, just as living as memory" (De Kerkhove, *ibid.*).

The proto-examples of the "body icon" can be identified in classical Greek sculpture towards the end of the 5th century BC. In those years, the Greece of city-states, inside its new political space won in the Persian wars, needed new monuments to glorify the memory of the great deeds done in the historic conflict between Greeks and "barbarians", and of the common Hellenic denominator that unified the Greek cities against the Medes and the Persians. In the *Peloponnesian War*, Thucydides inaugurated the politics of memory that relegated Herodotus to "mythology", long before Aristotle's' criticism. Though it is true that Herodotus intended to undermine the prestige of Homeric epic and the Trojan War by giving Greece something worthy of memory, this is not the reason behind his breaking of the ties with traditional memory. Indeed, this rhapsodic researcher of Halicarnassus never fails to pause before those who, like Darius, wish to leave a unique and unforgettable monument of themselves. These may be mainly material monuments, such as the iron spit of the courtesan Rodopi deposited near the temple of Delphi (2, 135), the channel dug out by Xerxes in the viscera of Mount Athos (7, 24), or the labyrinth built under the rule of twelve Egyptian kings near Lake Meri(2, 148). These are monuments that are recounted, recited as important undertakings, marvellous feats: things that are worthy of memory caught and secured by the eye that scrutinises and explores them, and by the ear that registers their fame and their praise.

Thucydides, on the contrary, is resolutely engaged in writing, or rather, in conceptual writing: what allows us to "see clearly", "the acquisition for eternity". His attention is not directed at a public of occasional listeners, but rather to readers situated outside of time and its normal surprises, in other words, a reader before the mirror of a completed, definitive and unchangeable book. According to Thucydides, the truth of "efficient" and "useful" history is a written history. But it is also memory protected from "rumours", purified of the falsity of what is merely overheard, rescued from the temptation of the pleasures of listening and storytelling. In the statue ofa hero or a god, this memory, which suffocates and kills private and local living memories, locates the "body icon's"duplicity as artificial creation, suspended half-way between the ecstatic movement of exceptional behaviour–the plane of thought of the "specialists of the soul"—and the level, easy journey of mobility doomed to oblivion—"the trick of the trade" that deceives us by "making most things seem real" (Dissoi Logoi, 3, 19, ap. FVS, II: 410).

Eikon is the technical term for a figurative, sculpted or drawn representation; it is the image created by the painter or sculptor. It is what seems to indicate the expression that Michele Psellos (P.ereg. daim., 821 Migne (PG, t. CXXII)) attributed to the poet Simonide di Chios: "The word is the image of reality". One of the salient aspects of the "new direction" introduced by classic statuary is the appearance of a signature at the base of the statue or inside the painting. In the relationship created by the artist with his figurative work, he discovers himself as agent, as creator half-way between reality and image[24].

The "body icon" is the cornerstone of this mutation, which turns the meaning of the statue upside down just as much as it does the traditional relationship the artist has with his work. It marks the epical passage from the sung, rhythmic word that comprised the oral traditions of archaic Greece from the 12th ^{to} the 9th centuries to the "word-dialogue" that organised the rational thought of the City[25] from the end of the 5th century. The "body icon" is an invention of writing. It emerges when the written sign immobilises the flow of the living word whose only law is to repeat itself in an infinite series of variations. In this sense, the classical statue[26]is an *éikon*: a figurative sign that substitutes the reality of human nature, transcending its physical limitations.

The ambivalence between the human and nature had already come to the surface in classical Greek sculpture, oscillating between conquest and

---

[24] Treu (1955:298) marks this change in direction in Hesiod's "Shield", in which he sees the "desire for illusionism" characterised by the term *Eikon,* as *Apáte* – "illusionary art". For an entire tradition, painting is an illusionary art, a "trick". The author of the *Dissoi Logoi* defines it as "an art where the best one is the one who is best at deceiving (*exapate*), making most things similar to reality" (Dissoi Logoi, 3, 10, ap. FSV, II: 410).

[25] Various studies, especially those of Louis Gernet and Jean Paul Vernant, have shown that the passage from oral memory to *logos* was a process of secularisation of the way of thinking that took place in the course of the 7th and 6th centuries B.C. in institutional practices, such as juridical and political practices. It is in social life that the conceptual framework and cognitive techniques that tend to favour the disappearance of so-called rational thought are simultaneously formed. The secularisation of the word occurs within this general framework, in which the social and the mental continually interfere.

[26] The first hundred years of Greek sculpture, or in other words the Archaic Period, dates from the end of the 6th century until the end of the Persian Wars in 480 B.C. The sculptures of the first seventy-five years (625-550 B.C.) have a limited number of models, of which the standing nude ephebe is the most common.

fusion, and between self-centred vision and self-centred orientation. In the 1980s, this would foster the main thread of the "triad" of "the symbolic, the imaginary and the real"[27], efficaciously highlighted by Lacan in *The Name of the Father* (2006). The Greek statue is in fact mediator and guarantor of temporal continuity, and so of the cultural coherence and the identity of the Athenian community, the *pólis* par excellence. Using the Olympic gods, and unbeknown to them, as it were, the city inventors created citizens, divinities that governed the city pantheon, gods closely involved in the daily dimension of social and political life. In the smooth, perfect bodies of Venus de Milo, Praxiteles' Hermes, Myron's *Dicobolus*, Polyclitus' *Doryphorus* and the *Nike* of Samothrace, an asymmetry between changes and survivals was produced.This happened in a culture in which scrutiny of old and new information was, yes, carried out by the memory of every individual, but always in the service and under the control of the social life of the *pólis*.

Nobody will deny that from the 6th to the 4th century there was a continuous buzz of gossip in the Greek house, where Xenophanes' curses, bellowed from the rooftop, echoed the indignant grumbling unhurriedly gathered by Plato the procurator as he moved from one room to the next, before reading his famousaddress. Halfway through the 4th century, as far as the Greeks were concerned the "master of all arts" continued to be Homer, whom the rhapsodists recited in the squares and whose works continued to be the cornerstone of a traditional aural system. In Plato's political project, from his *Republic* to his *Laws*, the identity of the immemorial voice that permeates the City's life acquired great evidence: its objective becomes that of converting this sea of voices into the service of the City. The classical statue is in fact made to "civilise" natural chaos, bringing it within the measure of the City, "at the frontier where the uncultivated and the civilised are next to one another in opposition, certainly, but also with reciprocal interpenetration" (Vernant, 1987:21).

It is at this frontier that Oedipus' wandering came to an end when, on leaving King Polybus' palace, he went to question the oracle of Delphi. The "frontier" is the narrow stretch where the roads leading to Megas,

---

[27] The source of inspiration for the Lacanian triad can be found in Lévi-Strauss' *L'éfficacité symbolique,* published in 1949. In this, the author proposes the succinct but unprecedented definition of an empty subconscious, with no content, pure organ of the symbolic function, which imposes laws of structure on a material of inarticulate elements that come from both reality and the reserve of images that we have all accumulated. The concept of the "individual myth" that was put forward in these same pages was taken from Lacan (1979) "*Le mythe individuel du névrosé*", "Ornicar", 1979, n. 17/18: 952.

from Thebes and from Daulia, meet to form the only route to the sanctuary at Delphi. It is at this crossroads, marked by two tall stones, that Oedipus encountered his father Laius' chariot with the herald Polyphonte, and killed them. At the crossroads of death, the tragedy of philosophy was consummated.Here, in the sanctuary of Delphi, the Word, the sacred, knowledge, self-consciousness; there, on the other side of the gulf, the body, otherness, blindness and enigma.

A climate of scandal, then, that situates, in the statue, the opposition between the *Republic* and the violence of a philosophy that has only just learned to express itself.Enclosed in an epic style as its primordial language, tradition petrifies itself in paradigms; transformed into statues, it offers the spectacle of socially acceptable models of behaviour. In the statue, the root of modern thought was already *in nuce*. Nature has no sense, if not in opposition to human works, whether they be called "culture", "society" or "history" in the language of philosophy and the social sciences; or "anthropized space", "technical mediation" or "oecumene" in more specialised terminology. The classical statue emerges as a new figure of the comparison whereby visual experience etches out the practice of writing in order to define the ambiguous limits of the *logos*: between writing and telling, between "inventing stories" and explaining tradition in a written space that still closely reflects oral memory. The Greek statue already testifies to the separation that occurs between the citizen of the *pólis* and the "uncultivated world": "wild animals, plants, uncultivated lands and youths in the sense that they are not yet civilised or integrated into civilisation (Vernant, 1987:20)".

Greek sculpture incorporates, in a form, the exchange between the world of the body and the world of places, inaugurating the idea of art as the middle term between the sensible and the ideal, capable presenting the highest things by making them appear to the senses.When Phidias finished the statue of Zeus in Olympia, he was worried that the statue might seem slanderous to the king of the gods and so implored him to give some sign of recognition. According to the legend, a bolt of lightning immediately struck the place where the statue had been placed. Zeus had triumphantly applauded his chosen artist and, with him, had blessed Greek civilisation for obeying the sovereign perfection of form.

Figure 3-5 (left): The Venus of Milos (Louvre Museum)
Figure 3-6 (right): Hermes of Olympia (Olympia Museum)

The Greek statue, whether sculpted from Parian marble or bronze, is now unquestionably associated with a *reflective attitude*, with a sharp loss of immediacy, which involves both the artist's self-awareness and the public's competence. It also leaves decisive tracesin the work itself, reflexively regulating its symbolic function[28] . Appearance becomes truth, and the surface allows the depths of the body to emerge as scenario of the"primary emotions"[29].

Emotions are largely sub-conscious processes whose first purpose is to put us into a state of alert, and prepare us to react with certain behaviours to particular environmental stimuli. They constitute the fundamental ways whereby consciousness emerges from the interactions between neuro-physiological processes and social environments, and they directly influence the general functions of the brain and of the organism, from the regulation of physiological states to thought processes (Damasio 1994; Spence et al., 1996). Over the years numerous studies have confirmed the existence of these retroactive mechanisms on the organism's responses to

[28] The classical mimetic paradigm interprets the image as a copy of real things: an ontologically degraded simulacrum, according to Plato, an instrument of active recognition of the world, according to Aristotle.
[29] Cf. Spiegel (2001: 125). For further information, see note 5, chap. 2.

emotional experience.For many researchers in this field, this constitutes a fundamental aspect of emotive processes. Some aspects of our evaluation system are innate, such as those linked to motivational systems of attachment and the search for novelty, while others are acquired through experience. The search for proximity and eye-contact with the care giver are not learned behaviours; rather, they are imprinted in the child's brain from the moment of birth[30].Genetic factors obviously play a fundamental role in determining the ways in which our evaluation system organises our behaviours in order to increase our chances of survival. Action, learning and development can be seen as phenomena strictly inter-connected throughout the course of our entire existence.

The reflective attitude as "debt to the world" is situated within the system of evaluation of the meanings that organise social behaviour on the basis of profiles of emotional activation.In the absence of such evaluation mechanisms, all stimuli from the outside world, as well as internally generated representation and states, would be perceived and elaborated in the same way. The temporal element of elaboration, its procedural and interminable character, strikes us here with absolute clarity. The authenticated image, the one that allows the technical prosthetics of sensibility to proceed gropingly towards an unknown "outside" infested withsimulacra, is therefore not only an image aware of its inevitable shortcomings. It is also an image *woven with time* and exposed, like the "flesh" named by Merleau-Ponty, tothe effects of reflective splitting that comprise the realm of time, *to the oscillation between before and after, between presence andmemory.*According to Aby Warburg (Gombrich, 1970: 283-306), the image resembles a coagulation of "mnemonic waves" (*mnemischen Wellen*) that can be retraced in their origins and their pathways. Only with this method, using the historian as an ultra-sensitive seismographer, can an image be read (in truth, any text), understanding its every internal vibration and every meaning.

---

30 On this topic, Bowlby hypothesises that every individual has an innate motivational system which is present from birth, which he calls the *attachment system.* Its purpose is to maintain the balance between the individual's internal safety conditions and external conditions. This balance is the outcome of the environment in which the individual finds himself. The mechanism is based on retroaction, or in other words on mechanisms of *feedback and forward,* and it is what keeps in balance the tendency to explore the physical and social environment in the absence of danger, and readiness to continue to explore when the environmental circumstances change again, giving the system the green light that says "danger has passed".

Thus, for Christian culture during the Middle Ages, mobility is an aspect of the transitory, the unstable (a good example is the theme of the wheel of fortune), the earthly and history. It characterises the man of flesh, the temptation of sin and the agitation of vice, which is contrasts withto the regular, celestial motion of unchangeable cycles, and even with the total absence of motion. What modernity adds to the Aristolean interpretation, radically modyfing it, is the prerequisite of autonomy. At the dawn of modernity, this was exactly the way that an artist-scientist like Leonardo conceived the work of the image on the intelligibility of the physical world and its hidden regularity. Precisely in that it calls attention to itself and its independence from the rule of mere reproduction, the "image-screen" can exploit its capacity to *actively perlustrate* the world and hold in reserve heuristic potentialities not expressed in the course of time. This bipolar opening, which keeps in constant (but also unstable and indeterminate) relation the constructive and the referential work of the image, has for a long time been an inexhaustible, creative resource in modern art. In his last works, Maurice Merleau-Ponty (1979, 2003) brought to light the reflexivity of the body (or the "flesh" as he calls it).This reflexivity, according to Merleau-Ponty, reaches its maximum deployment in *vision,* and finds its exemplary referent in the image.The eye that explores the world is what it is only insofar as it is also part of it, as both the "subject" and the "object" of vision. Which, as Merleau-Ponty goes on to note, gives a structural, even prior to a narrative, dimension to the frequent appearance of mirrors and other optical or specular devices in carnal space.

Using the Freudian expression (1975) *Durcharbeitung*,[31] the "body icon" constitutes the area of intersection between memory (the archive and its unconscious areas) and the repetition compulsion. In other words, the body icon's way to actualise itself becomes an explicit game between different media forms which make themselves felt in their opacity, in order to solicit the spectator to reflect on the irreducible "otherness" of whatever is being represented, on its irreducible *being different* (Bolter and Grusin, 1999: 73). By the "otherness of what is represented" is meant the coherent and historically datable form that brings together, in the concreteness of a

[31] Starting from his original designation of the term *–Durcharbeitung-* Freud means the long passage of working-through that occupies the area of intersection between memory (the archive and the unconscious zones) and its pathogenic way of realising itself. As is well-known, grief, losing someone dear or an object of affection, are worked through. In his work *Grief and Melancholy*, Freud (1976) maintained that grief is not a *state,* but a *process,* governed by the *realityprinciple.*

specific context, "mnestic traces: the sensorial and corporeal shreds"[32] in which Freud recognised the fragmented character of memory. Such "forms" are "negatives" (Artieri, 2009:71) and they providethe premises for a single living being to "think differently" through a plurality of scenes of an imaginary life.

Let us consider the following samples of post-modern "body icons": *Spider Man*, the famous comics icon;Madonna, an equally famous icon in the world of show-business; and the Jamaican athlete, Usain Bolt, the most sensational icon of a man who goes "beyond the limits" of the body.

1. Spiderman

Spider man is the icon that made Marvel Comics the privileged partner of American film studios. The origins of the transformation of the clumsy teenager, Peter Parker, into a New York super-hero is etched on the marble of the walls of college and high-school culture. While visiting a natural history museum, Parker is bitten by a spider that transmits most of the natural traits of a spider to him: speed, muscle tone, shock resistance, and the ability to weave a web that binds his instinctive body to the context of action. One morning, after being bitten, Parker wakes up and contemplates his body, which has suddenly become like a sculpture. He shuts himself

[32]A significant example of these "forms" are, for example in Proust's work, the disconnected paved streets (that remind him of his trip to Venice), the stiff napkins (the trip to Balbec) and the famous taste of the *madeleine* (his youth in Combray). These memories, which for Freud represent the primary scene and for Proust are "involuntary" memories, entwine past and present in the immediate context, which the "body icon" then organises, creating itself as incredible, liminal "viaticum" of human inter-subjectivity.

away in his room to practise the fist-movement that releases the sticky white substance his body has started to secrete. Parker only uses this power, otherwise difficult to control, to win over Mary Jane, “the girl next door”.

2. And Then There Was Madonna...

We can use George-Claude Guilbert’s words in *Madonna as Post-Modern Myth: “How One Star’s Self-construction Rewrites Sex, Gender and the American dream”* (2002: 158).

The personage of Madonna incessantly plays on the meanings of virgin, saint, mother, vamp and femme fatale. It is infinitely multiplied as Nordic Goddess, great Mediterranean whore, sadomasochistic priestess, gay idol, manipulator and business woman. An ongoing palimpsest, she never ceases to rise from her own ashes and her every rebirth becomes an important international happening. The real queen of pop, nobody else has become a living icon as she has. It doesn’t matter whether she is singing, dancing or acting. What will remain is her “body icon”, with its infinite “looks”.

But what exactly does Madonna do? She sings. But it is an over-simplification to define the fifteenth episode of *Glee* as a tribute to Madonna, or to say that the music written in the last thirty years by the “queen of Pop” has inspired a consolidated phenomenon. The truth is that when Madonna hits *Glee* she creates the same results as a category-4 tornado, both during and after. This interesting and triumphant interpretation was brought to this episode by director and screenwriter Ryan Murphy. At first it caused many of the series’ fans to turn up their noses, for it was thought that the idea of focusing an episode on a single artist with only one style of music would make the weekly episodes of *Glee* monotonous and complacent. And that if the artist in question was Madonna, the risk of becoming dull and predictable would be very high. But Murphy is surprising. He doesn’t limit the outcomes, he doesn’t take *Glee* away from its usual context. He makes shrewd and unexpected choices out of the singer’s enormous repertoire, once again making the most of the opportunity to transmit a fundamental message clearly.

Like Madonna (though the original title *The Power Madonna* is better chosen) is an anthem to women and the important role they play in society.

2 E FU MADONNA...

www.queenmadonna.com

Utilizziamo le parole di Gorge-Claude Guilbert in Madonna as Postmodern Myth: How One Star's Self-construction Rewrites Sex, Gender and the American Dream (2002: 158):
Il suo personaggio gioca in modo incessante con i significati vergine, santa, madre, vamp, donna fatale, si moltiplica all'infinito: dea nordica o gran puttana mediterranea, sacerdotessa sadomaso, idolo gay, manipolatrice, donna d'affari, palinsesto costante, non cessa mai di riprodursi dalle proprie ceneri e ogni sua rinascita è un evento di portata internazionale.Nessuna più di lei, vera regina del pop, ha saputo diventare icona, in vita. Che canti, balli o reciti, è ininfluente. Quel che rimarrà è il suo "corpo-icona" dagli infiniti look. Ma che cosa fa, esattamente, Madonna? Canta. È abbastanza riduttivo definire il quindicesimo episodio di Glee come un tributo a Madonna, o sostene-re che la musica creata negli ultimi trent'anni dalla "Regina del Pop" abbia ispirato un fenomeno già consolidato.

In realtà Madonna si abbatte su Glee creando, sia durante sia dopo, lo stesso risultato che avrebbe avuto un tornado di quarta categoria. È una chiave di lettura interessante e vincente quella che Ryan Murphy – regista e sceneg-giatore della puntata – ha dato a un episodio che in un primo momen-to aveva fatto storcere il naso a più di un fan della serie. Si pensava che l'idea di incentrare un episodio su un unico artista e un solo genere musicale, avrebbe reso l'appuntamento settimanale con Glee monotono e compiacente.
Se poi l'artista in questione è Madonna, il rischio di cadere nel banale e nello scontato era molto alto. Invece Murphy sorprende. Non dosa per nulla gli sviluppi, non estrapola Glee dal contesto in cui si muove, fa scelte sapienti e non ovvie nell'enorme repertorio della cantante, e ancora una volta sfrutta l'occasione per trasmettere in modo chiaro un messaggio fondamentale.
Come Madonna (ma è più azzeccato il titolo originale The Power Madonna) è un inno alle donne e al ruolo importantissimo che esse hanno nella società.

3. Usain Bolt, Beyond the CT

Jamaica: the seed of speed grows here, between sugar canes and cacao bushes. An indolent, tropical forest—rum, yam, bananas and lots of Superman plantations. Trelawny County, the same address for Usain Bolt, Ben Johnson and Veronica Campbell Brown, all athletes who burn away the seconds of the straight line. Usain Bolt is the fastest man ever. He is the current Olympic and World champion in the 100 meters, 200 meters and the 4X 100 meters relay, specialities in which he also won world records in the Olympic Games of Berlin (for the 4X 100m) and the World Games of Berlin in 2009 (for the 100m and the 200m). His great gifts as a sprinter have earned him the nickname of Lightning Bolt.

This is what makes the world's fastest man: an early start to the day, a spontaneous smile, supple muscles, correct nutrition, a sound nervous system, a technique that aids without flattering, feet of steel and a body-form capable of attaining bio-mechanical levels never before been reached.

The "body icons" considered above are *intermedial products* (Montani, 2010: XIII). They are based on the assumption that only by starting from an active comparison *between* the image's various technical formats (for instance, optical and digital) and its various conversational forms (fictional and documentary, for example) can we do justice to the evidence of media or other sources of facts. This technical and what could be called "exteriorised" comprehension of the image loses any residual trace of obscurity if referred to the radical and unattainable "off-screen" persona incorporated in the "body icon". As observed by Giddes, 1990), modern communication technologies are in the early stages of the process of dis-embedding or de-contextualisation. By such means social relations are extracted or estranged from a context of local interaction, to be reformulated in an undetermined spatial-temporal sphere. In its specific acting the "body icon" determines, by contrast, the contextual limit that will become the habitat of its power over the spectator's emotions, obviously bearing in mind Bolter and Grusin's description(2002: 23):

> a small spectator who watches the videotape of Batman wearing Batman's hat, eating a take-away meal wrapped in Batman paper and playing with a Batman toy.

In their various costumes, each of the "body icons" considered above is a sort of artificial "glorious body" which, from his or her muscles to the empathy they inspire, forms the basis of the Personage's leadership. Leaders are people who prevail and distinguish themselves, whether it be in war, politics orbusiness, as well as in religion or show business. But why do others, the people who do not command but obey, or at least follow, choose that specific person as a leader andreference point? As the Dutch psychologists von Vugt and Auja (2010) demonstrated, the first important factor in prehistoric times was, obviously, physical presence, height and muscle, given that every type of problem was solved with force. A second leadership factor lies in belonging to a clan, party or religion, as various studies show. With the passage from hunting to agriculture, there appears another reason for leadership, destined to become ever more important: wealth. The farmers who accumulated the most grain and later the most animals, the most goods with which they could obtain other goods, realised how easy it was to command within their community.

But it is not physical strength, or belonging, or even wealth that creates the"glorious bodies" of the western god or hero. What is, instead, the difference between them, as illustrated by the "body icons" studied above, and normal spectators is that they are capable of dominating a context served by technology, using the power brought into play by their primary emotions. Parker's Spiderman galvanizes audiences because the spider's bite he received at the New York Natural History Museum gives him manual dexterity, the security of a panoramic view of things, and the effectiveness of movement, through which he acquires, much as Ulysses with the Trojan horse, sovereign power over his primary emotion on the border of the event taking place on the rooftops of Manhattan. Madonna revolutionizes the story line, transforming characters by offering them a new philosophy on life, and allows the cardinal element of the show, music, to cross the limits of the music-room to invade and involve the whole McKinley High School in a collective, colourful euphoria of extraordinary choreography. First shyly, almost as being forced to do so. Then in an almost uncontrolled fashion, so that anyone who has a feeling for song and dance feels an irresistible urge to crash through the screen in order to join Finn, Rachel, Kurt, Mercedes and all the other Glee Club kids. Usain Bolt, on the other hand, is a highly significant champion of the "extreme use of the body" based on a combination of the ancestral power of his memory of belonging, his natural mastery of his body, and hisunrelenting daily training.Through this training he develops strength and flexibility as well as specific psychophysical skills (such as concentration, management of energy, and the pleasure of fatigue).

As the champions studied show, this is the dawning of a new world that the "body icon" encourages to manifest itself. It is a world that is regenerated as an unlimitedhorizon of reference, and produces the discovery of the body's hidden resources. The "body icon" is without a doubt a special "creation". Making something special implies intent, and the intent here is to distinguish an object or action from the ordinary, evoking emotions through rhythm, design and colour. Making something special is a behaviour that augments group cohesion and thus could lead to an advantage for survival. Dissanayake (1988) suggests that numerous factors play a role in the motivations of art—manipulation, perception, emotion, symbolism and cognition. These developed alongside other human characteristics like tool-making, the need for order, language, the making of symbols, self-consciousness, the creation of culture, sociability and adaptability. But while all of these are motivated by what was needed for biologic fitness in our ancestral environment (food, successful reproduction, security and awareness of predators, friendship and status), the "body icon" presents itself, by contrast, as the source of a limited repertoire of stereotyped behaviours. Thus it confirms the principle behind Wilson's (quoted in Carrol, 2007:31) penetrating explanation of the adaptive function of the arts:

> There was not enough time for human heredity to meet the vastness of contingent possibilities revealed by evolutionary intelligence [....] The arts have filled this void.

3.4 The Secret Way: the Rhythmic Body

The body was theoretically invented in the 20th century. This invention derived from psychoanalysis, when Freud, observing Charcot's exhibition of bodies at Salpetrière, identified conversion hysteria and recognized it as the starting point from which all his research would stem: the subconscious speaks through the body. This was a decisive first step, since it opened up the question of somatisation and led to the consideration of body image in the formation of the subject, later known as "skin-ego[33]". Husserl brought the next important contribution by focusing on the body as the "cradle" of all meaning. His influence was particularly felt in France, and led to Merleau-Ponty's idea of body as "embodied consciousness" and "cornerstone of the world" (1945: 97). The third stage of this discovery of the body came out of Marcel Mauss's surprise during WW I at seeing the British infantry march with a different step from the French. The notion of "body technique" has had a profound influence on all historical, anthropological and sociological thinking about the body.

Yet, despite Mauss's enlightening intuition, more often plundered than fully understood, the body was relegated to the subconscious, stowed away inside the "subject" and included as the "own body" among the social icons of culture. It was as if knowing, in the most general sense, meant "representing oneself", according to the famous philosophical theory that contrasts the immediate data of consciousness to the principal elements of representation. On the basis of this theory, culture is a social substitute for instinctual automatism, which plays a role of symbolic mediation. Indeed, language, representations of reality, mythological tales, religion, artistic expression, technology, scientific knowledge, philosophy, legal systems and behavioural models constitute the "forms" we use to mediate our relationship with our own "self", with others and with things (Crespi, 1996: 14). As he sometimes does with real things, modern man assigns symbolic, verbal or mental behaviours to these "substitutes", according to the paradigm that considers man's "own self" as *outside of him* from a developmental point of view: a technical world, originally the result of the body's externalization, that takes on increasing importance to the point of manifesting itself as a universe endowed with a life of its own. This poses the embarrassing problem of determining which part is played by the inner man (his faculties) and which by the outer one (technical prosthetics). Thus the body, the place of the human by definition, becomes the "object" in which a transformation is revealed and effected. On the one hand this transformation seems to want to dispossess the person from his territory – corporeality—and make him rest on virtuality; on the other hand, and not only in our day, it changes his character into forms which bring us to the post-human and the trans-human (the latter being a term that probably dates back to a piece written by John Huxley in 1927).

---

[33] Cf. Anzieu D. (1985), *Le Moi-Peau* (1985), Paris, Dunod; Bick E., "The Experience of the Skin in Early Object Relations", International Journal of Psychoanalysis, 1968, 49: 558–566

The charismatic vector of this transformation is "one's own body". In the 5th *Cartesian Meditation*, Husserl[34] attributes the connection between the self and the world to this notion. "One's own body" manifests itself as an "ego-logical" property functional to "the social reproduction of the self". Fundamentally, its character is *subjective* in that it is the structure in which consciousness[35] appears. Tendentially solipsistic by nature, it is based on the "methodological exclusion of the other's subjectivity" (described by Husserl himself as a form of abstract art) and it takes place in the first person. The "other" is, in fact, a "body like mine" (an "almost own body"): "what my body could be if it were there and not here".

The "own body" governs our linguistic codes, our semantics and our social syntaxes, generating in the 20th century the phenomenon of corporeism: a neologism mainly used to indicate the discourses and practices that give prominence to the body as investment in values; examples are the fields of literature, philosophy, psycho-sociological, analysis and therapy, as well as those of the plastic arts, sports and intermedial imagery. Of these, the *social media* deserve a particular mention, in that they are practices with "shared" content "from the grassroots". This can be seen at work today, in the sharing of images and videos on platforms like Flickr and You Tube, in the activities of users of social network sites like Facebook, in online worlds and in blogs or more traditional forums.

Through our "own bodies" we have become a being watched on "the world's stage". The medium *par excellenc*e of this stage is the icon, whose main role is to highlight social temporalities such as biographical trajectories enveloped in impermanence, self-referential episodes isolated from historical space and time,"parallel and complementary" universes (Mazzoli, 2009: 9), and communications platforms which interpret collective emotional intelligence as a playful pact "beyond politics and culture" (Susca, in Jenkins, 2007: 364).

The gazeis formed precisely in the mobile zone, which is undefined and can be revoked at any point of the meeting between the activity of perception and the passiveness of the perceived. (Dinoi, 2008). The force

---

[34] Cf. Husserl E. (1994), *Méditations cartesiennes* (1931), Paris, PUF ; Merleau-Ponty M. (1942), *La Structure du comportement*, Paris, PUF ; (1945) *Phénoménologie de la Perception*, Paris, Gallimard ; Henry, M. (1990), «Pathos avec», in Phénoménplogie matérielle, Paris, PUF.

[35] Cf. Schutz A., Luckman Th. (1973), *The Structures of Life-World*, Heinemann, London; Luckmann Th. (1993), *Remarks on Personal Identity: Inner, Social, and Historical Time,* in A. Jacobson-Wigging (ed. by), *Identity: Personal and Socio-cultural. A Symposium*, Almqwist & Wiksell, Uppsala.

of the icon lies in certifying, by means of media wrapping, opacities of sense that preserve important elements of meaning of the entire world of acting and suffering that have not yet been "processed" by the prosthetic techniques of sensibility.

In the body become icon, we find ourselves facing a bona fide industry of the wonderful. What we have to pay attention to is the fact that in the "body-icon" we are in the presence of vigorous social tendencies that entail a new type of annihilation of the biological time and temporal rhythms which have governed our species since its earliest phylogenetic origins. Times blur, creating an eternal universe that is not in expansion, but which is automatically conserving itself. It is not cyclical, but it is unpredictable; and it is neither recurrent nor sequential.Associated with an electronically integrated multi-mediasystem, the icon shapes social temporality into a sequence of moments, each of which is a pleasurable flash. It is the media time of zapping, replay and surfing: a "point of being" (De Kerkhove, 2003: 26) that guarantees multiple, simultaneous and instant access to every point in the world and shapes the map-like a suspension of everything we can know and feel: what the so-called New Architects[36] and a great many architectural and design theoreticians call a "skin".

As in Alice's wonderland, "instantaneous time" is the small, hidden door that, over the various stages of the "secret story" of the human body, has allowed it to actively produce the meanings which in our day find, in goods and cultural products, a tool for the incessant creation of an orderly world endowed with sense as dictated by the key elements of our culture. These are: appearance as the dominant category of reality, the myth of physical form, contempt for routineand daily labour, artistic activity as the only fulfilment worthy of man and, more generally, the desire that many have to shut their eyes to what is happening and show us a fake world that is colourful and happy, made in their image and likeness.

In its ungraspable otherness, the icon weaves a dreamlike design that expands to become the *stade du miroir*, the "mirror state" (Lacan, 1953: 159). Through the gaze it organizes the horizon of reference of the processes of mass interaction.From the analyses of the French sociologist, Maffesoli – especially his most recent ones (2009) - the icon emerges as:

---

[36] Stephen Perella, Brian Massum, Marcos Novak, Lars Spuybroek (cf. De Kerkhove, D.) (2003), The Conquest of time.Society and Democracy in the Age of Network.*La conquista del tempo. Società e democrazia nell'era della rete*, Rome, Editori Riuniti.

> ...the almost-resigned transfer of the ideological and socio-political stereotypes of our time to the screen of cultural meta-history, in an attempt to regain the detachment of iconological vision from actual reality, albeit in a problematic way, convinced that myths [are] an oxymoron: "the dark clarity that acts as "headlight" (Puglisi, *Introduction* to Maffesoli, 2009: 10).

Unlike Barthes' (1993: XVI) hermeneutical research, in which the icon answers a need to capture the immediacy of the message from the proposition ofthe image, for Maffesoli (2009: 29) it is an "imagining form" capable of rousing "everyday passions and emotions"—"the reasons of the heart that the reason does not know". One of the slogans linked to the resounding success of the recent film "Avatar" is that "you don't watch Avatar, you live it". This is a perfect example of the statement quoted above. According to the analysis of the French sociologist, the icon is a "place in chiaroscuro" (Le Quéau, 2007) that "burns the soul of the world", as James Joyce wrote about the Ulysses who is its eternal creature.

Along the "tenuous catwalk of the present" (Kundera, 1983), what scintillates today is the wonderful saga of Harry Potter, the new "no future" Dioniso of lingerie, the Brazilian Chaball, the "human beast"; Hermès or "the soul of things"; the android Sarkolène, the emotional amplifier of Interet blogs; and Che Guevara, the angel of death of the Cuban revolution.

Like a "guardian angel" of human emotions, each and every icon – and the ones already mentioned represent significant examples – embodies, according to Maffesoli (2009: 20) "the emergence of the secret king that governs its inmost essence", namely the body, rooted in biological rhythms both individual and species-related, which is definitively fading away in the exhibition of its icons.

Figure 3-7: Harrypotter.warnerbros.it/hp7a/

The human body has an intrinsic and constitutive spatial-temporal structure. If it is able to inhabit and understand space-time– and not simply be there – it is because the body itself constitutes this space-time in act and is self-aware. Dan Lloyd (2006:312), a philosopher at Trinity College in Hartford, Connecticut and author of a fascinating book entitled *Radiant Cool: A Novel Theory of Consciousness*,has this to say about the relationship between body and time:

> Temporality is the skeleton that sustains objects and objective reality. Once we understand temporality, the way to the embodiment of consciousness opens wide before us.

The body is where time, corporeality and context intersect. The adaptive experience of the individual and the species develops by means of the *chrono-biological* maps, which orient its development and design its functionality on the basis of the fabric of relationships that make up the specificity of a given context. The formative process, in fact, springs from the synergy of various interacting dynamics, in which genetic reality and epigenetic development are entwined, thanks to the constant correlation of the biological dimension with relational and socio-cultural matrixes. There is a close and rigorous interdependence between time and body, which is rooted in the life of cells and molecules. For as long as we are and continue to be a psychosomatic unit, we will be possessed by the temporal tension that always accompanies our thoughts, works, days, our lives and our deaths. What gives us the history of the body, its density and temporal resonance, is material experience itself. The ultimate originality of this experience lies in its being the meeting point for individual and social experience and for individual reference and the collective norm.

As Damasio rightly points out, concerning the orientation of temporal corporeality:

> The whole construction of consciousness, from its simplest to its most complex forms, from the non-verbal consciousness of images to literary and verbal consciousness, depends on the ability to map what happens over time inside our organism, around our organism, to our organism and with our organism—one thing after another, causing yet another thing, continuing ad infinitum.

Time resides inside the human body and "in the vortex of its own adventures" it produces "a wise contingency" (Mazzarella, 2004, 160). According to Biuso (2009:142), body and time constitute "the score for the concert of consciousness". On the basis of that definition, the body is the outcome of the metabolism that transforms the other into self and uses oxygen, herbs and the meat of other animals to form the thinking substance of the present. Animated life is constituted by radical motility. The metabolism's need for food and other resources drives animals to move around in larger or smaller areas, and even cover continental distances. As a species, *Homo sapiens* exhibits this trait in every climate and in the most varied of habitats, even hostile ones. Moving around in a space cannot be separated from movement in time. To move from one point in a space to another does mean merely to occupy different places, it is not limited to topology but is always a chronobiology as well. On the basis of this diachronic and synchronic development, the accumulation of human knowledge is carried out where networks, cognition and powers cross inside the human body.

Like all living organisms, the human body is also organized in a specific temporal structure in which all vital functions show variability over periods of time that range from a few milliseconds to several months. In man, as in less evolved organisms, many biological rhythms follow the frequencies of periodic stimuli that come from the outside environment, while many others are determined by internal pacemakers that act independently of any environmental input. In fact, external influences are not simply superimposed, but are, in turn, modulated by the rhythms dictated by endogenous pacemakers. With regard to time, the variability of physiological condition also leads to the body's differing "susceptibility" to disease. On the other hand, even pathogenetic mechanisms can be irregular over time, both as regards their presence and their severity, and this means that their "suitability" to produce clinical manifestations of diseasesvaries over time.

We can define the rhythmic body concisely as the ongoing result of the history of the organism, and of its interactions with the environment, in the construction of identity and of an individual's characteristics. In other words, the rhythmic body can be considered as the whole of the chronobiological maps that at any given moment characterize the organization of a living system as a network of sequentially produced events, within which each event is generated by the transformation of the previous state into the following one. Through the rhythmic body, every kind of knowledge is born of interaction with the *outside,* with what is not "my body". Everyone's body is the physical place from which the world originates and expands, at the meeting point of our skin, the antenna that captures what is happening around us, starting from our awareness of space and time.

We create reality by reacting to real events, both internal and external, but our brains are not limited to mirroring these events. Instead, it elaborates them according to its own models. Much of this brain activity is unconscious. So for us reality is neither subjective nor objective. It is a construction of material connections that mingle what happens in the internal and external physical world, by inserting information acquired through experience into the circuits of our brains. This happens by means of correspondences created by neural connections over time, between the characteristics of events and the catalogue ofreactions available to the brain so that it can carry out its regulatory function. These correspondences are not fixed. They can be manipulated by our bodies in relation to connections with environmental objects. Every different set of connections creates new experiences. The rhythmic body is, in fact, *the source of activity* of our conscious. Starting with the general aim of survival and well-being, a particular corporeal manipulation incorporates the emotions, feelings and reasoning that define the way we live. The exact modalities by which activating profiles are "connected" constitute a key problem for scientists studying the brain, the mind and society. The question is how simultaneously activated processes are connected to each other to give rise to "*continuity ofexperience*". Here we put forward the hypothesis that, through the interactive sets that produce objects of use and tools, the topographical construction of the rhythmic body regulates the order of selective processes within which biological rhythms evolve in the environment, thereby giving rise to the continuity of a given experience.Proof of this theory can be found in cultures with an oral tradition. Leroi-Gourhan has shown that the itineraries of the rhythmic body are what create traditional "domesticity" as a result of the progressive equilibrium which man's body is able to negotiate with the environmental cycles of the habitat in which his social life is concentrated. In the practices of this body in different contexts it becomes our nature to *generate culture*.

The rhythmic body, in fact, is a direction-finder, like a map or a compass, which expresses itself not just with language, but also in the material stratification of places, in the spaces that have been inhabited for centuries and millennia, in urban structures. Starting with the Palaeolithic revolution – but even before that, with the use of tools – the modern human brain has been irrevocably influenced by the interaction between the body and material and symbolic artefacts. In their paper the *Philosophy of Mind*: *Classical and Contemporary Readings*, philosophers Andy Clark and David Chalmers (2002) put forward an "embodied" vision of the relationship between mind and outside world that caused much debate then and is still topical today. These two philosophers advanced the concept of *active externalism,* based on the active role played by the environment in guiding cognitive processes. In the cases they examined, the human organism is connected to an external entity by a bi-directional interaction, which results in an associated system that can be viewed as a cognitive system in the strict sense of the term. All components of the system play an active role and together they govern behaviour, acting in the same way as cognition usually does. If we take away the external element the operational capacity of the system will decrease, just as if we had taken away a part of its brain. The theory put forward by the two philosophers is being validated today by a growing number of studies in cognitive science, in areas as diverse as the theory of situated cognition (Suchman, 1987), the study of robotics applied to the real world (Beer, 1989), dynamic approaches to infant development (Thelen and Smith, 1994) and research into the cognitive properties of collective agents, (Hutchins, 1995).

Thus, to consider cognition as being "extensive" introduces a noticeable change to the methodology of scientific investigations. In practice, explicative methods once deemed appropriate only in the analysis of "inner" processes are now being adapted to studying the exterior, giving rise to the hope that our understanding of cognition will be enhanced. For some, this type of externalism is hard to digest. One reason for this might be that many identify cognition with consciousness, and to them it seems implausible that consciousness enlarges itself outside the head. But not all cognitive processes are conscious, at least not in standard usage. It is widely accepted that beyond the limits of awareness processes of all types play a crucial role in the cognitive process; for example, in recovering memory, in linguistic processes and in acquiring skills. The biological brain evolved and matured insofar as the presence of an external environment subject to manipulation could be relied on. It would certainly seem that evolution has favoured internal competences that are triggered in order to make use of the world environment to rearrange memory load and even to change the nature of computational problems themselves. Our sensorial apparatuses have evolved by trusting the environment in various ways; by exploiting contingent facts relative to the structure of natural scenarios, for example (Ullman and Richards, 1984), or by taking advantageof computational shortcuts produced by bodily movements and locomotion (Blake and Youille, 1992). Once the fundamental role of the environment in conditioning the evolution and development of cognition has been recognised, extended cognition becomes a true cognitive process. Within that process, the rhythmic body becomes the central point in which time, action and context intersect. The final result is, in fact, what Bergson and Gehlen describe as "action": *the concrete and intense act of interacting, combining procedural memory with the practice of contingency.*

How can the flow of time be fixed, how can its significant stages be "tamed" and how can the sign system of its salient moments be designed? In the repertoires of social anthropology and ethnography (written, film and web graphics) the history of archaic humanity provides many significant answers to these questions. The work of socially constructing temporality undertaken by very different systems and societies shows that time can be rendered familiar, accessible and manageable rather than foreign, incomprehensible and hostile. The paradox of time and logos are the same: that is to say, the paradox of speech that claims to name being. It is the truth of the *vox publica* which, extensively echoing and sounding out speaking subjects, constitutes them as *vox universalis*, as public *selves*, internalised intersubjectivity, "an I that is an us and an us that is an I" as Hegel said, and Husserl repeated in his own way in his 5th *Cartesian Mediation*. De Saussure (2009) believes that the signs of language take on the features of time because he does not see the eye with which tradition, and he himself, sees and measures language. In other words, he does not realise that it is the notion of time that takes on the character of language, understood as conventional public sign, the linear alphabet of the west, precise sequence (counterpoint): in other words, speech. In the involuntary conclusion that Saussure reached (which can also be illustrated by some emblematic passages from Kant's transcendental "*Aesthetics*", where the depiction of time is significantly entrusted to the line), the world of the epistemic *ratio* is summarized as the whole of the imaginable worlds that Cacciari (1985) defined in his *Icons of the Law* as "reversals" – "going backwards, almost like music! – of the arrow of time, instantaneous times, stratifications of time happening simultaneously".

From Bergson to Husserl, we have witnessed the claim of the conscious, or the internal experience of time as concrete, lived time, in contrast to the spatialized, instrumental and abstract time of science. But duration, flow, current and such like are, in turn, no other than metaphors for the discursive *logos,* masks of the psychically internalized *vox publica.*The otherness of the world is the inescapable condition of reference (as well as of the creativity of semiosis), but it would not be attestable without a reflexive and referential attestation that takes into consideration the way linguistic forms shape the world and save it from its differences.

What Maffesoli (2009: 148) metaphorically defines as the "genome of human nature", referring to "return to animality in the human" (*ibidem*, 52) caused by fashion, theatricality and the game of appearances, translates, in the body, as a return to the limelight of a "ritualized savage state, that kind of internal and external savagery that lies at the origin of all conquests" (*ibidem*). Our theological, philosophical and political traditions have always defined man by contrasting him with "the animal": "with that part ofthe self, or that area of humanity, that is preventively bestialized" (Esposito, 2007: 182).

Just like the "imagining forms" presented by Maffesoli, "the return to the animal" promises the "conquest of time" as a volatile intertwining of the real and the virtual. Not to wish for a "miracle", but that the magic of memory – (to remain with Harry Potter) will still work a little, and that people will not see the extent of the damage.

Iconicity is contagious. If visible, an icon has a contaminating effect on the words used to describe it. If verbal, it will contaminate the images we want to use to represent it. Today the techniques of iconicity are, more than ever,fundamental organs of communicative reproduction. , and the destiny of our most tangible reality is at risk from those techniques. In these techniques the fate of our most tangible reality is at stake: in other words, the *rhythmic concatenation* between our biological bodies and the environment that surrounds us, referring both to the temporal depth of specification and its effective contextual function. This *rhythmic body*—which "creates relationships between entities that are completely different, like a man, an animal, a microorganism, but also a tree, a season or an atmosphere" – constitutes the "difference that separates the modern western world from all peoples, past and present, who have not deemed it necessary to proceed with naturalizing the world" (Descola (2005: 57). In western culture, the "body's secret history" is the iconic confirmation of this difference.

CHAPTER FOUR

ICONIC SPACE MAPS FROM THE ITALIAN RENAISSANCE TO "ELECTRONIC AUTISM"

4.1. The "Upside Down World" of Modern Culture

In light of the process described as the convergence of modes[1]that is reshaping digital networking, the post-modern paradigm gives way to the analysis of the assortment of technologies, devices and applications that support the process of cultural mutation our world is experiencing.

What is it that drives us to devote ourselves so diligently – almost as if obliged to do so – to all the various social networks, on line videos, blogs, digital radio, media activism and Wikileaks? Perhaps we ought to stop and consider the repercussions wrought by social networks on our bodies, which are already overloaded with information.

The devaluation of the body to icon and its inextricable relationship with social and technological activities make up the argument of this chapter. This chapter is dedicated to explaining why the move away from the old world of hypertext towards social media is accompanied by the disruptive adoption of a cultural system designed as an illusionary construct with a substitutive function. In the history of the image, the transition from analogical to digital paved the way to accessing the immaterial, and the computerised image itself became immaterial. The whole ontological relationship that had devalued and dramatized our dialogue with appearances since the days of the ancient Greeks has been turned on its head. Representation is on the way out, heading towards the final chapter of the long metamorphosis in which things increasingly seem to be a pale copy of images.*Here we have the visual.*

Today the dimension of caring, which for Heidegger amply defined men's being in the world and their responsibility to the phenomenon of

---

[1] Jenkins (2006: 3) wrote, "... the most important dimension of communication divergence occurs within the brains of individual consumers and through their social interaction with others".

existence itself, seems to be reduced to the narcissistic cult of the body imagemanifested in "post-modern objects" and "machines beyond the body". These have taken on, in a variety of different styles of seduction, the cultural privilege once belonging to modernity. The desire to change the skin we inhabit has become "territorialized" – as in Almodóvar's film – on screens, digital sites and social networks. There has been a shuffling of the cards in an attempt to dance over the ruins of passing time. Everything is in play, everything is art[2]. Madonna like Michelangelo. Comic strips like an eight-hundred page book by Tolstoy. The already advanced transformation of man into spectator is up for discussion. The multiplication of screens creates a "visual party".Even tragedy is turned into entertainment and overwhelms perceptions, steals emotions, seizes events and urges us to participate with sympathetic indifference.

In pre-modern times the body was there "for the long run"; it controlled the space and time of an experience, and through "technical acting" it created the rules of connection and inclusion that organize a territory. The arrival of modernity meant that the body became the mobile epicentre of an infinite variety of tensions, each time representing the gap between the forces in play at any given moment in time and space. But what is at the centre of all this?In the words of Deleuze-Guattari (1972:45), "At the centre is the desiring-machine, the celibate machine of eternal return".

Commenting on Durkheim, Moscovici (1988: 74) speaks of a "resurrection of the imago[3]" that acts profoundly on the social body. In an impulse of reversibility, images live again and regenerate the social body—all of society or the small tribal group that acts as support. Maffesoli (1995: X) describes this as the "imaginary world" or the "imaginative power of reality". It is possible to propose a classification of these images to evaluate social and institutional reactions and therefore the propensity of the "rule" to take over the body and the processes it generates.The simulation of the old digital world that is "deaf, blind and dumb" allows nerves and muscles to relax.Paradoxically, the aesthetic dimension seems

---

[2] Post-modern art, not yet defined as such, started up in the 60s, when media images became a taking-off point for artworks. Warhol revisited Hollywood icons, and Lichtenstein cartoons took Duchamp's message to the extreme. "We are all Duchamp's grandchildren" could easily be their slogan -- a way to repeat the past with a clean conscience.

[3] In psychoanalysis, *imago* is the term used to describe an image with strong affective value which in early childhood represented the persons the child was physically closest to, and which are later projected onto real or imaginary people who are important in one's life.

to protect the human body from its openness to the unpredictability of experience as it develops *within* and *beyond* the "flow of space" (Castells, 1996: 473). As Montani (2009: 13) rightly observed, "there can be no doubt that the advent of 'bio-aesthetics[4]'is imminent, characterized by a disquieting levelling, reduction and channelling of emotions". The self-referentiality of all forms of representation is increasing, and exponentially so, accompanied by a corresponding specialization of the *aesthesis* that receives it. The following question brings this complex problem closer to home: how helpful are current theoretical paradigms in analysing what concerns above all the equilibrium of the human body in the *habitat* in which social life is concentrated, including electronic spaces? And why did our culture—specifically, modern western European culture—choose at some point to go down the road that has led it to identify with a culture like the one we live in today, where *high fidelity images* have a role that may not be their only one, but is "certainly dominant"?

The increasingly sophisticated technologies used to represent the visible world have a history of their own.They are the product of a development that has being going on for more than five centuries, based on a single tactical move, a single *adequatio mentis et rei:* the accommodation of the mind, conceived of as a flat, perfectly dry slate to the thing, the face of the earth progressively reduced to the same scheme.

This stratagem dates back to at least as far as Aristotle's *Metaphysics* [I, 992b33] and *De Anima* [III, 4, 429]; therefore it came after the revolution in which the Athenian Cleisthenes, employing geometric measurement, first used the map as operative model of the city and the territory (Farinelli, 2003:160-61). In modern times, with Descartes, nature was definitively transformed into a geometric space, and matter into "pure spatiality" (Borkenau, 1934: 347-349). What is more, the "long chains of reason, totally simple and straightforward, habitually used by geometricians to carry out their most complicated demonstrations" (Descartes, 1637: 303) became "the method", a term meaning literally "what comes after the concrete journey and both determines and directs it". If it is true that we have never been modern, it is not, as Bruno Latour (1991: 38) maintains, because modernity is based on the schism between nature and culture. Actually, the opposite is true: the program of modernity depends first of all on the attempt to use a single, coherent set of rules as the foundation of our knowledge of matter and of mind itself, uniting them under the aegis of the logic incorporated in Euclidean geometrical extension.

[4] See: Esposito R. (2004), *Bios.Biopolitica e filosofia*, Turin,Einaudi, Par.1.3, 1.4.

The crucial shift in the history of modern Europe consists exactly in this: the transformation of the world into a series of models that are limited but open, in that they constantly refer to something that is *absent* because it is *external,* has made possible the "*space of representation*" guaranteed on paper. In the words of Lefebvre (1974: 59), this is a space which "the imagination tries to modify and occupy". While for pre-modern societies space constituted the *anchor of time*, in modern societies it levels time by standardizing its social organization.This change, according to Giddens (1990b), coincides with the expansion of modernity.

Quattrocento Florentine perspective was instrumental in establishing the new space of representation. It came into being in Florence, beneath the Portico degli Innocenti, which Brunelleschi built sometime between 1419 and 1422, thereby inventing, with Florentine perspective, the Renaissance. Favouring sight as a cognitive tool over all the other senses and forcing the observer to stand still, it creates an illusion that rewrites the statute on which the subject-world relationship – and therefore the very nature of the subject – is based. This veritable cultural revolution, consolidated by the mid-16th century, consists in the transition from a direct relationship with reality to a systematic and pervasive relationship with its image: in other words, in the substitution of what exists with its representation, with absolute equivalence between the world and its representation. As to painting, it was Etienne Gilson (1958: 148-149, 257 ff.) who came up with an inspired explanation: "whereas before Giotto a painting was a thing, an object, from Giotto to Cézanne it becomes the portrayal of a thing".

Panofsky (1961: 37-38) calls Florentine linear perspective "artificial perspective", the preferred vector of the transformation of territory into space, first in Europe and then, thanks to colonization, in the rest of the world. It was also Panofsky who quite unintentionally explained the limitations of this transformation, which in his view remains ideal, but instead forcefully influences not just perception and representation, but also the actual construction of reality in modern terms. Thus, Panofsky explains that the modern construction of perspective is in total contrast to our psycho-physiological apparatus, which has no perception of continuity, uniformity or isotropy. This is because our sight is always limited, and so the fundamental relations of the organization of its parts (in front-behind, above-below, left-right) generally have different values. On the contrary, modern perspective is based on the metric space of Euclidean geometry, or better on Ptolemy's cartographic principle (Edgerton, 1975), and consequently produces a continuous, uniform and isotropic expanse. All the points within this expanse are without content and are therefore simple expressions of ideal relationships. From each of these points, equal constructions towards every other point and in all directions are possible, which does not occur in reality. In short, this is a mathematical space. Technically speaking, linear perspective is the beginning of the in-depth order of a conventional space.

As Fracastel (1957: 64) observes:

> The basis of Quattrocento representation of space is the concept of man who, conscious by then of his autonomy and proud of his ability to distinguish, moved in a new space, known from that moment onwards as the "theatre of the world".

When in the 15th century linear perspective was established as the spatial agent of modernity, man's experience became the "*theatre of vision*". In his "Treatise on Painting", Leonardo da Vinci explained very clearly that painting is the unifying perspective and, touching on all fields of study, provides the thread that joins them all. His approach to scientific knowledge was *visual*. It was a painter's approach: "Painting embraces within itself all the forms of nature" These words of Leonardo provide the key to understanding his science. He repeatedly asserted – especially in his early writings – that painting involves studying all natural forms, and he also emphasized the close connection between the artistic representation of those forms and the intellectual understanding of their intrinsic nature and their underlying theoretical principles[5]. Thanks to his genial dissection of bodies, da Vinci for the first time uncovered the complete path of visual signals from the pupil and crystalline lens through the optic nerve to a particular cavity in the brain, known to neurologists today as the "third ventricle". He considered these discoveries about the optics and physiology of vision as the basis of the science of painting, starting with the extraordinary innovation of perspective.

If there is one of Leonardo's works that has ended up as a symbol of our times, it has to be the *Vitruvian Man*[6], which he drew in the early

---

[5] When studying the proportions of the human body, Leonardo compared them to Renaissance buildings. His research on muscles and bones led him to study and draw gears and levers, thereby creating a connection between animal physiology and engineering. He started with the human eye, and his in-depth studies of the anatomy of the eye and the origins of vision had no equal in his times. He paid particular attention to the connections between the eye and the brain, which he illustrated in a series of drawings of the human skull.

[6] This famous drawing is preserved in the Cabinet of Drawings and Prints in the Academy Gallery in Venice. Today the drawing, intended to show the centrality of man, is one that all Italians, and most Europeans, carry around with them every day in their pockets, as it is the symbol found on the one-euro coin. Da Vinci drew

1490s. One could in fact say that this date, which virtually coincides with the discovery of America, corresponds perfectly with the intimate nature of an image that saw the light at the dawning of the modern age and in a certain sense evokes future expectations. Of this epical change (and it might seem strange to be able to call it that since we are, after all, talking about a work not much more than twenty by thirty centimetres—to be precise, 34.3 x 24.5), the work shows an indelible mark in the artistic choices made by Leonardo from conception to completion. The image epitomizes these choices so well that—pardon a possibly audacious literary comparison—if Pico della Mirandola's *De Hominis dignitate,* published in 1496, has been rightly considered the "manifesto" of the Renaissance, Leonardo's drawing is its figurative equivalent.

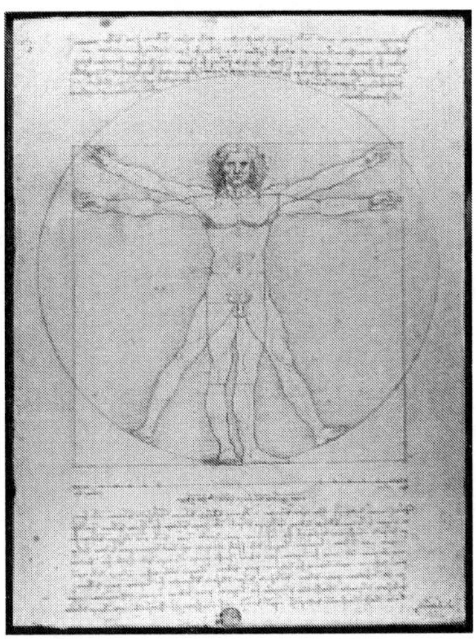

Figure 4-1: Leonardo da Vinci's Vitruvian Man

---

it in 1490, revisiting the text in the third book of *De Architectura* by the Roman architect, Vitruvius, which treats the proportions of the human body. In describing the drawing he had just completed, da Vinci wrote, "The architect Vetruvius writes in his book on architecture that the measurements of man are in nature distributed in this manner. The centre of the extended limbs is the navel; in fact, if a man lies on his back with his arms and legs spread apart and you place a compass in his umbilicus and draw a circle, the tips of his fingers on his hands and his toes on his feet will touch it". His man, standing tall with arms and legs spread apart, can be exactly retraced in the geometric figures considered as perfect, the circle and the square.

The technique of linear perspective derives from the inventory of forms and symbolic relationships expressed by Renaissance plastic space, often incorrectly codified in the formula of "universal man". Accepting this theory means admitting that the outside world is an object that men have viewed in the same way throughout the ages and in all countries, and that a discovery made during a particular era, the Renaissance, was the key that could unlock all the secrets of the universe by analysing structures so privileged as to represent, after the Christian revelation, a sort of new natural revelation, one that time had shown to be more or less in agreement with the earlier one.

The study of space in the Renaissance leads us to take another look at some opinions on what this period was like for the people who were alive then, and also what place it deserves among the great periods of history. The inventors of perspective representation of space are "creators of illusion," and not more or less sagacious imitators of reality. This new space is a mixture of geometry and symbolic representation, in which technical knowledge exists to serve individual and collective ideas. Renaissance plastic space is no more than the perfect expression of the image of the world belonging to a particular group of men. The first speculations of the Renaissance were based on the metric evaluation of the diaphanous aspects of the universe. They were founded on a new evaluation of the relationship with objects, and involved an image of the outside world seen as a positive reality distinct from man and God, called Nature. The compromise that was made between line and colour was based on two arbitrary conventions: the absolutely unreal convention of a single source of light, and the equally unreal reduction of vision to monocular perspective.

The trick behind Florentine perspective lies in the assumption that the dimensions of objects depend on their distance from the observer. Florentine art, in contrast to ancient art, did not consider the angle of vision to be relevant, despite that fact that the retina is curved, not flat (Panofsky, 39-43). But this is a fiction which, in favouring the eye as a cognitive tool over all the other senses and forcing the observer to be immobile, changes the statute which governs the relationship between the subject and the world, and consequently the nature of the subject itself. Samuel Edgerton (1975:30-40) noted that this formulation of perspective was a "visual metaphor" of the whole Florentine world at that time: the Medici oligarchy dominated politics, and banking and commerce were being increasingly rationalised, based on mathematical regularity and the use of double-entry book keeping. The Tuscan hills were terraced, planted with elegant rows of olive trees and parallel rows of vines, all overseen by a central land office. Proportion and regularity were appreciated in all areas of culture, and it was expected that they would also govern social decorum and form.

Despite some occasional variations or intentional violations, the rules of perspective dominated the representation of space in art through to the twentieth century. Then, the influence of the impressionists and of Cézanne and the Cubists shattered the world of perspective, as if an earthquake had devastated the reticular precision of the pavements of a Renaissance street. When the impressionists abandoned their studios and started to paint in the open air, they discovered new points of view, shades of colour and types of light. They broke Alberti's rule, according to which the canvas was to be positioned exactly one meter from the background and directly in front of the subject, by choosing to set it up anywhere and everywhere, even in eccentric positions, so as to create new compositions. In fact, the impressionists moved in and around the setting they were painting, and the frame was no longer a proscenium of a cubic section of space, as it had been in the past.

Still, however much the field and the angle of impressionist space might vary, it remained essentially a space looked at from a specific point of view.The cubists' assault on the closed form was the most significant change that characterized the beginning of the 20th century. They discovered that they could, and should, deform objects and change colours only in consideration of their artistic sensibilities.

In 1905, Hugo von Hofmannsthal wrote that in his time nature was "multiplicity and indeterminacy", and that what other generations believed to be steady was in fact slippery. X rays could pass through the human body and other material screens.Boccioni wondered who could still believe in the opacity of the human body and compared the penetrative capacity ofx-rays to the refined and varied sensibilities of Futuristic art. Thomas Mann investigated the impact of the x-ray in *The Magic Mountain*: when shown by the doctor what his hand looked like on a fluoroscopic screen, seeing how the x-ray "disintegrated, annihilated and dissolved the flesh of his living body", for the first time in his life Hans understood that he was going to die. Einstein's theory of relativity questioned the stability of all spatially extended forms[7]. Under these

[7] Einstein introduced this particular theory in an article entitled *The Electrodynamics of Moving Bodies* (1905), in which he maintained that bodies change shape, when they move in respect to a fixed reference system. The general theory of relativity demolished conventional understanding of the stability of the entire material universe. Classical physics taught that all bodies are elastically deformable and that changes in temperature alter their volume. According to

circumstances, the grid of a system with Cartesian co-ordinates is useless to trace the plane of movement[8].

Among all the changes we have mentioned, Cézanne's was the first to introduce a truly heterogeneous space in a single painting. Using multiple perspectives of the same object, he dispensed with linear perspective, violated perspective in landscapes, and tried to reconcile the properties of volumes in a three-dimensional space with the two-dimensionality of a pictorial plane.The Cubists repeatedly expressed their debt to Cézanne and used his techniques to create even more radical treatments of space. Their use of multiple perspectives shows a strong resemblance to cinema, which dissolved the uniformity of visual space. The two pioneers of Cubism, Picasso and Braque, integrated Cézanne's innovations with those of cinema. In so doing, they brought about the most significant revolution in the rendering of pictorial space since the 15th century: they abandoned the uniform space of linear perspective, painting objects in multiple spaces and from multiple perspectives, with visions of interiors similar to those made possible by x-rays.

Like modern art, cinema offered new possibilities to modify space. A theatre audience watched what was happening within a determined framework, from a single viewpoint and from a fixed distance, in a space that did not change from the time the curtain went up to when it went down. In films, however, space could be manipulated in many ways. A shot could be changed by moving the camera, and visual space could move continually in a panoramic shot. The multiplicity of spaces that camera techniques could produce was further extended during the editing process, which made rapid shifts from one viewpoint to another possible, and even more the dissolution of spatial coherence. The possibility of changing the way a shot was framed suggested ways to break up the traditional fixed proscenium scene, while in the 1890s the change from gas to electric light made it possible to light different parts of the set in a single scene and to vary the colour and intensity of light, creating changing chiaroscuro effects.

Einstein, every fragment of matter in the universe generates a gravitational force that accelerates all material fields in its field and modifies their visible dimension. Thus there are no completely rigid bodies.

[8] The ordered world of geometry, shown graphically by distinctly squared Cartesian co-ordinates, became complex and unstable and could only be represented with respect to a system of reference; in this case "from a matrix of slimy molluscs of reference that were constantly changing their form when accelerated by masses and by the variable distances of countless moving particles" (Lorentz, 1942: 48).

With cinema the statute of the image changed and *fact became artefact*. Cinema brought about two extraordinary contributions to the sense of distance in visual arts: the close-up and fast cutting. Close-ups create intimacy, while fast cutting creates a dramatic sense of separation and distance. In an article written in 1907, Remy de Gourmont, the French critic and novelist, announced his passion for cinema, which allowed him to "travel around the world" and to experience in an instant the difference between desert and ocean, unexpected juxtapositions of completely different environments with all the associated emotions. The cinema was a huge "magic lantern" that showed the most remote and inaccessible corners of the world.

Cinema's visual power comes from the big screen, grand actions and great stories, narrated so as to make spectators "feel big" as they unfold by appealing to their *primary emotions*. The power of film, as Ricciardi (2009:23) commented, is "the power of suggestion" and "the sensational force that comes from the spectator's identifying with what is happening on the screen". In essence, a moving image has the power to capture an audience, to keep an entire group of people completely engrossed.

From the early 1990s, television, the *archetype* of mass communications, developed new methods of transmission – cable and satellite – determining the code for reading the events written about, negotiated, represented and celebrated on the small screen.According to the theory developed by Dyan and Katz in *Media Events*: *The Live Broadcasting of History* (1992), live broadcasts of "historic" events constitute the greatest moment of celebration in mass communication. Seen live, thanks to the power of the media, a real event becomes spectacle, causing the breakdown of the physical nature of place and of the body's experience in a territory. The most incisive confirmations of this "spectatorialising" come in our times from computerized graphics, especially if we think of the latest developments of this art form, which are designed to create simulated reality.

Ours has been defined as a "civilisation of images". The definition would be even more accurate if we were to add that it is the civilisation in which a particular type of image - specifically *trompe-l'oeil*- has become extraordinarily realistic, thanks to new production technologies and thespread ofthe iconic. The invention of the photograph made this clear, and the advent of film and television even more so. The world and its representation are dragged into an extra-rapid exposition; *the very instant that time is fixed in space*. "Here" and "now" are united, in photography, in a whole not unlike a flash of light that, paradoxically, crystallizes the transitory. This is what Benjamin defines as the "expositive value of a work of art" (1936: 28).[9] This thesis first introduced the "media" interpretation of technology, highlighting the repercussions it would have on our perceptive practices.This is what he wrote:

> With the various methods of technical reproduction of a work of art, its opportunities to be exhibited have increased to such an extent that the quantitative discrepancy between its two poles has become, as in prehistoric times, a qualitative transformation of its nature. (...) In prehistoric times, through the absolute importance of its cult value, the work of art had become an instrument of magic, and in a certain sense only later came to be recognized as a work of art. In the same way, today the absolute importance given to its exhibition value makes the work of art a formation with entirely new functions, among which the one we are conscious of, the artistic function, may be recognized as marginal in the future. This much is certain: today photography and cinema offer the most fertile occasions for the recognition of this new fact.

As the expository value of a work became increasingly important, its cultural foundation was replaced with what Benjamin called its "political" foundation, by which he meant that a work could be reintroduced into the fabric of public space, thanks to the dissolution of connection to tradition and to a peculiar detachment of the work from itself. According to Benjamin, the temporalised history of the modern is in fact an attempt to grasp—as a *continuum* of the course of time—the intermittent and occasional consciousness of the irrevocability of a past fixed by the image-memory. To make what is transitional essential, and to fix the irrevocable, is the secret of aesthetic time.Aesthetic time is the place of memories and of appearances which, in disguised or counterfeited form, can also be concealed in the temporalisation of history in modernity.

Thus, from the shadow of the reciprocal disguise in which art and technology seem spellbound in modernity, Benjamin allows a glimmer of light to shine through and reveal the ineliminable distance that separates aesthetic and historic time. In his *Gesammelte Schriften* (1974, vol.I, t. I: 1247) he writes: "The past can be fixed only as an image that flashes once and for ever in the instant when it can be known. Its authenticity comes from its transient nature".

This *transience of the new* is appearance that objectivises itself, making itself available and instrumental in technology. The fundamental

---

[9] Benjamin maintains that the emergence of the technology of photographic and phonographic reproducibility led to the appearance of expressive forms originating primarily in a mechanical process. That process was from the start, by its very nature, oriented towards reproducibility. Cinema is the classic example.

direction of Benjamin's research during the most productive years of his work was the analysis of tradition and its interruption. In this connection, Szondi has made a very perspicacious observation (*Aut aut*, 1982: 190):

> Unlike Proust, Benjamin does not flee from the future. On the contrary, he seeks it in the emotional turmoil of the childhood experiences whose traumas caused him to go, as it were, into hibernation; whereas he had to enter into the present as though into his tomb. Benjamin's "lost time" is not the past, it is the future.

The time of expectation and desire that mere chronological growth is not yet able to meet has found shelter in childhood traumas. Benjamin establishes an affinity between distant past and messianic future, the archaic and the *novissimum* (the most recent). This means that the past preserves itself as if it had been impressed on a photographic plate; however it only shows its image at a determined "hour", which depends on its relationship to other images that have also revealed themselves. But not even transmission of the past is certain; there is no past that can call itself "safe forever in the granaries of the present" (Benjamin, 1966: 90). In this sense, the real historic present is nothing more than the past of stored images that develop at exactly the same time and remain temporarily immobile on the "threshold of time": "modernity always summons up the prehistoric past". (Benjamin, 1966:150). The claim of the aesthetic time of the image against the temporalisation of history in modernity developed in Benjamin as a result of his awareness that this temporalisation is nothing more than a falsification of original time. Here lies the deep-rooted sense of the attention he paid to modernity as a primarily aesthetic phenomenon. His scrutiny is born of awareness that modern historic time will only allow itself to be interpreted by art, understood as an *artificial construction of transience*.

For Benjamin, the aesthetic image, the time of music, of the idea oflanguage and of the *mémoire involontarie,* represent the counter-movement that breaks the spell of externalised history. Historic time can return to its origins, that is to say "a birth of becoming and passing" (Benjamin, *ibidem*: 226). It is snatched from the brink of a blind temporal *continuum* only by a memory that is able to save it, as irrevocable past, in an image[10]. In the ashes of the past and the primitive, of fairy tales and

---

[10] The purpose of allegories, the pre-eminent Baroque form of expression, was to represent "illuminated oblivion", though not to the point of creating complete clarity – examples include the allusive and perturbing and petrified walled-in past, along with cadavers and ruins. In these conceptual images of the primitive, the

involuntary memories, the embers of the future are buried and preserved. When the past joins the future, there is a flash of "now-time", of the instant when the future returns from its exile in the past and settles in the historic present. In that paradoxical moment, appearance and disappearance short-circuit the notion of "duration" and become images of "exquisite cadavers". (The *cadavre exquis* is part of surrealist poetry). Proust's *Madeleine's* andGozzano's "good things in very bad taste" are children of photography, cinema and metropolitan life. There can be no nostalgia without modernity or even, as the sociologist Fred Davis maintains, "nostalgia exists for the media, thanks to the media and in the media". And yet, there has been another change since the days of Proust and silent films. There has, in fact, been another decisive stage, and modern nostalgia has been followed by post-modern nostalgia. Modern nostalgia rues the loss of youth and secretly thinks that there was a time when things were better. Today's neo-nostalgia, which fuels nostalgia for the past – *retro*– should perhaps be renamed. *Vintage* might be its right name: being a word associated with fashion, it gives the idea that the past is no longer a continuous journey but a *wardrobe* in which all periods are present at the same time, in the form of media images. So: *vintage* is a sort of ready-to-wear past, and therefore the opposite of history and perhaps even of memory. The alarming thing, however, is that what is felt to be one's most personal and secret essence – one's body rhythm – is something quite ordinary, produced in series. It is like the scene in *Blade Runner*, when the android discovers that what she had thought were her childhood memories were actually created by a multinational business.

4.2 The Screen and the Mirror: the Electronic "*Bricoleur*"

Is it possible to express real feeling in a simulated space? Today this pressing question concerns the protocols of imagination that comprise the social interaction, empathic drives, passions and aesthetic concepts of the new media, generating the illusion of "liberating" the body from its spatial conditioning. This follows the post-modern assumption[11] that the "realty

---

theories of Bachofen and Klages on pre-history and chthonics merge with those of Freud about the workings of the psyche, in which time is a force-field of contrasting motives (one timeless and the other experiencing time), which results in time as conservation and repetition of the past, despite the passage of years and decades.

[11] Postmodernism appeared for the first time as a philosophical term in 1979 in a book entitled *La condition postmoderne* by the French intellectual, Jean Francois Lyotard, in which he affirms that different groups of people use the same language

of the world" is created by putting "the manifold narrations deriving from, and shaped by, our imagination" (Vattimo, 1989:55) into a given context.

The term "imaginary" owes much to the analysis of the imagination carried out by Sartre, Lacan and Bachelard, especially where their investigations are focused on the myths and symbols that make up the genetic heritage of the forms of representation of a given social system. In this context, we cannot fail to refer to Henry Corbin's metaphysics of the imagination[12], Paul Ricoeur's insightful research[13] and, of course, *La poétique de la rêverie,* which is a sort of condensation of Gaston Bachelard's legacy[14]. For Bachelard, imagining a movement calls to mind the concept of imagining something material. In the volume *L'air e les songes* (1943), Bachelard suggests that the axes of the imagination's fundamental intentions are the routes followed by the most important gestures of the human animal towards his natural surrounding environment, which is directly prolonged by *Homo faber's* primitive technological and social institutions. But this direction is reversible, because the elementary environment shows evidence of the attitude adopted towards hardness, fluidity and fire. We could say that every gesture calls up its own kind of material and looks for its own tool, and that all materials extracted from the surrounding environment, and all tools, represent the sign of a "lapsed gesture".

This reciprocal genesis of gesture and environment, symbolized by the hearth, has been highlighted in American social psychology. Kardiner, for example (1968), places it within the notions of "primarity" and "secondarity", pointing to the "on this side" and "on that side" of the basic personality. According to Kardiner, when individuals and their drives receive a normative imprint from the surrounding environment, they respond in turn, as a "secondary effect", making profound modifications in the material environment and on institutions. Bastide concludes a detailed study by demonstrating the pivotal role played by society as regards the *libido.* Individual drives always have a "social bed" which it can be easy to

---

differently and so see the world from very different personal viewpoints. Consequently, the idea of a single predominant vision vanishes completely. Lyotard goes on to say that this leads to the deduction that all interpretations are convincing and all are on the same plane. According to him, this convergence of interpretations constitutes the essence of postmodernism.

[12] Corbin, H. (1959*), L'imagination créatrice dans le soufisme d'Ibn Arabi*, Paris, Flammarion.

[13] Ricoeur, P. (1960), *Finitude et culpabilitè*, II, *La symbolique du mal*, Paris, Aubier

[14] Bachelard G. (1960), *La poétique de la rêverie*, Paris, P.U.F.

penetrate, or which on the contrary can throw up obstacles, so that "the protective system of the libido is not a pure creation of the individual, or a personal mythology."

It is precisely in this encounter that the "cultural complexes" (Bachelard 1943: 26) that come to substitute psychoanalytic complexes are formed. Inside the emergent configurations of these complexes and their organization are situated the empty zones, the non-codified or insufficiently codified margins that allow the individual or group "a frequently considerable [degree of] initiative" – so long as it can "resist the dizziness of the void" – as Gruzinski observes (1994: 362) when speaking about the indigenous societies of Spanish Mexico. The replacement of pictographic expression with alphabetic culture was far more than a mere question of translation or transcription. Indeed, it seems that the written use of pictographic information led necessarily to a significant loss, even more disturbing because it was irremediable and with no possibility of verbalization. This loss concerned the principles of both the selection and codification of the pertinent elements of the environment, going beyond the sphere of aesthetic categories in order to involve the implicit foundations of every construction of reality. Here we are probably touching on the deepest and least well-explored sediments of a culture, those which, while never made explicit, in fact underlie the specificity of cultural configuration: that which, in the *rhythmic body,* supports man's flesh-and-bone history and its transformations through the social use of technology.

Technology is the *condition for the existence of the species called Homo sapiens.* It gives expression to the exclusively human faculty of opening up *paths not predicted* in the normal course of nature. This radically innovative aspect is what the second chorus of *Antigone* records, in a tone of amazed concern. Singing about the power *hýper elpída* (beyond all hope) of writing, the tragic chorus documents its ambivalence towards the temporality it has introduced into the Greek *pólis*. Indeed, the appropriation of language through writing produces the problematic trait of technology in western culture: the performative self-reference that founds the *logical form of symbols.*[15]

The alphabet represented the fundamental principle of the translation of every human sense, and of every experience written about in books, and translated them into words, novels, theatre and poetry[16]. The translation of

---

[15] Here I deliberately repeat the expression Wittgenstein uses in the *Tractatus* (1921). To be exact, Wittgenstein distinguishes the *representative form* from the pure logical form, which is a sort of "form of the forms".

[16] As long as the text was associated with a physical thing, the crucial qualities attributed to it by writers and readers were "linearity, limitation and fixity" (Bolter,

sensory life into text is a translation of protection from real life, in which an experience is modified on the basis of a theory. The organization of the human "subject"– both the psychological and the political subject – depends on the portion of knowledge contained in writing, and in books. In the written text, the word functions as a barrier between sensory life and its exploration of the environment.

The first known testimony of the use of the word *textus* (no longer in the etymological sense of the weaving of a textile, but in the metaphorical sense) to distinguish the stable product of writing from the flow of words in oral discourse, is found in Quintilian's *Institutio oratoria*. This meaning of "text" already speaks of itself in the first person, communicating about its own representations, emotions and feelings—in general, about the inner psychological data inseparable from the "self"[17]. For semiologists, a piece of music is a text, as is an icon, a film a video, a poster[18]. The interest of semiology and semiotics is for all meaningful phenomenona, whether they belong to culture or characterise everyday activities. The idea, above all for the structuralists, is to found a science of language, signs and culture, based on the text.

When information technologies encountered the media languages of the 20th-century, the hypertext was invented[19]as an idea and project that preceded the computer. It comprises a body of material that is interconnected

---

J.D. 1991: 3).

[17] In the *Notebooks,* written prior to the *Tractatus* Wittgenstein affirms that "the self is the profound mystery" (Wittgenstein, 1921, 5.8.16) and in the *Tractatus* the question of the self is expressly re-proposed under the theme of solipsism (Wittgenstein, 1935: 85).

[18] The so-called linguistic turn originated in the Ferdinand de Saussure's university courses and in schools of linguistics and folklore studies, especially in eastern Europe, in Czechoslovakia and the Soviet Union. From the late 1950s, Paris became the capital of the various movements (structuralism, semiology, semiotics) associated with the linguistic turn.

[19] The history of the word "hypertext" is that of a technological phenomenon (the Web) and also of the combining of theoretical and applied research, of concepts and objects, and of the two cultures, humanistic and scientific. But it is also the history of a man, Ted Nelson, the father of the hypertext, who coined the word in 1965: He is the author of works that have had a profound influence in defining the discipline of computer science as the expression of a specific language and method of analysis and research. For further analysis , cf. Nelson, T.H (1972*),* "As We Will Think", in *Proceedings of Online 72 Conference*, Brunel University, Uxbridge; Nelson, T.H., "Replacing the Printed Word", in Lavington, S.H., ed., *Information Processing 80* (Proceedings of IFIP 80 World Computer Conference), North-Holland Publishing Co., pp. 1013–1023.

in such a complicated way, that it cannot be presented or represented on paper (Nelson, 1992: 1–17). Underlying this project there is the idea of literacy, "a system of interconnected ideas" which, according to Nelson, refers to any type of data that can be archived in any form whatsoever (be it writing, audio, image and video). Besides one's own name and the word "self", which distinguish the technology of writing, the hypertext inaugurates the postmodern poetics of the fragment[20].

The starting point of the hypertext is the Platonic questioning of vision. This questioning is based on the status of the image, and in particular, on the horizon of sense that founds the "destiny of the arts". The flexibility of alphabetic language translates the individual's sensory adaptation to the environment into aesthetic assemblage[21], in order to determine the specialization of the human body in its interaction with the world and the environment. Perceptual "commerce" with the sensible world constitutes the breadth of the aesthetic sphere as one of the deepest probes ever to have penetrated the subject of modernity. From this commerce, the invisible reverse side of what is available to sight, while soliciting sight to go beyond it, the body becomes the *bricoleur*[22]between the alphabet and the world. It is dematerialized, an *avatar*, a reflecting surface that can lead to the illusion of the double reality[23] that underlies the "heroic actions" of

---

[20] The hypertext aims at multi-sequential perception that proceeds by jumps. It expresses the necessity of a way of reading and an author-reader relationship that fosters the qualities of connection, ductility and openness already indicated by Ted Nelson in 1965.

[21] From the dawning of the Academy in ancient Greece there have been disputes about the laws and priorities governing the "sense of beauty". This search necessarily involved the attempt to describe the role of the body and its sensory apparatus. The word "aesthetics" derives from the Greek *aísthesis,* meaning "capacity to feel". It should be clear, however, that when we speak of aesthetics; we are dealing with a modern discipline. Kant, Schiller and Nietzsche created the lines along which it developed, constituting its "object" as the intimate exposition of human feeling to the sphere of the production of sense.

[22] An example of relevant works in this context, on the social studies of science, is foundin Knorr-Cetina, K.D., Mulkay, M. (ed. by) (1983), *Science Observed: Perspectives on the Social Studies of Science*; cfr. also Traweek, S. (1988), *Beamtimes and Lifetimes: the World of High Energy Physicists* , Cambridge Mass., Harvard University Press.

[23] For further development of this concept, cf. Baudrillard J. (1981), *Simulacres et Si- mulation*, Mayenne, Galileé; Perniola, M. (1990), *Virtualità e perfezione, Enigmi. Il momento egizio nella società e nell'arte.* Genova, Costa & Nolan.

fans, gamers and bloggers, who are like paradoxical "bodiless bodies"[24] roaming the maps of popular culture.[25]

The understanding of electronic space as a reality, substituting or parallel to "real reality", originated with the program of a group active in the 1990s in architecture and art schools in the United States and Canada. This group considered the interactive environment generated by the computer to be aesthetics, anthropology and "poetic thaumaturgy". In this period the use of the interface as a phantasmagorical game of the "human comedy" became popular in North America. The "magical" matrix was the term *cyberspace*, coined in 1982[26]by the science-fiction writer, William Gibson, to define an approach to the new informatics modes as a "web of connections among social, cultural and political institutions". Inside the space of this "web", the "hacker-hero" of this novel "dances" as he "dreams of swimming among the files" (Turkle, 1997: 41). Out of this matrix came the phantom of colourful neologisms – cyberspaceian, cyborg, cyber-punk, cybernaut, cyberdeck, cybersex, cybertech, cyber world, cyberentities(Benedikt, 1991) – that comprise the laboratory of Platonic metaphysics known as cyberspace. The "cowboy" of *Neuromancer* is no novelty in the history of western culture. "Never-seen-before worlds" in fact date back to Plato, and especially to neo-Platonism, if we consider the "disembodied human sensoria" which, according to the theoreticians of cyberspace, digital technologies have brought about and made independent of the physical body. As Maldonado (2007: 11) has rightly observed:

> This subject deserves careful analysis, in that it surreptitiously revives certain themes that have occupied the attention of philosophical thought for centuries, such as the problem of the real (or not real) existence of the material world, and the not very different one of the mind-matter relationship.

---

[24] Bodiless bodies" refers to considerable research sectors that have come out in the last decade, such as the literature on computer-mediated communication and online social environments, feminist cyberliterature and post-humanist literature (for closer examination, cf. Shilling, C. 2008, *Changing Bodies, Habit, Crisis and Creativity*, London, Sage).

[25] The expression Popular Culturedenotes the New Mediascene as a combination of cultural features and practices that portray the way individuals and societies relate to means of communication. Today the great majority of cultural products are not mass products. Indeed, we live in a world of infinite niches and sub-genres. (Cf. Jenkins, H. (2006), *Convergence Culture*, New York, New York University Press; Ital. ed., Jenkins, H: (2011), *Cultura convergente*, Milan, Apogeo.

[26]Gibson, W. (1984), *Neuromancer,* Ace, New York.

The paradoxical approach to the informatics culture is summarised in the so-called "aesthetics of simulation", which Sherry Turkle[27], a North-American psychoanalyst of the Lacanian school, launched at MIT in the 1980s. Using this term, she indicated the *compuscreen* (computer screen) as an "immaterial place for subjective dynamics". Employing the language of cinema, she proposed the small screen as a "relational artefact" centred on the subject's intimacy, privacy and "personalization". This journey "from the outer to the inner" treated the interface as a "theatre of humans" and the computer as a "machine for intimacy". As Turkle observed (1997: 27):

> When we are face to face with our image in the mirror of the machine we come to see ourselves differently: this is a strong signal of change, with attention to transformations of the spirit, of our depths and cultures, and not of economics or work or institutionalized politics.

Turkle coined the phrase "aesthetics of simulation" along the lines of the characterization of the postmodern that Frederic Jameson had already proposed in the 1984 article "Postmodernism or the Cultural Logic of Late Capitalism"[28]. In this article Jameson put forward a new model of spatial thought which would allow us to "register the complexity of our world". He suggested a new "aesthetics of cognitive mapping" which would give precedence to surface over depth, to simulation over reality, and to play over seriousness. The "soft approach" to electronic space became popular. The emerging culture of simulation encouraged ways of thinking that were non-hierarchical and based on negotiation. This so-called "kitchen mathematics", associated with situated learning[29], was founded on sensations and our everyday physical being. The computer screen was proposed as the "boundary between self and non-self, connected to the seductions of the interface". In simulation games, one enters the world behind the screen, like Alice passing through the looking glass.

---

[27] Professor of Social Studies of Science and Technology at MIT, founder and director of the MIT Initiative on Technology and Self. Cfr. Turkle, S. (1996), *Life on the Screen*, Cambridge, MA, The MIT Press; Turkle S. (2005), *The Second Self. Computers and the Human Spirit,* Cambridge MA, The MIT Press.

[28] Jameson, F. (July-August 1984), "Postmodernism, or the Cultural Logic of Late Capitalism", *New Left Review*, July-August 1984.

[29] An example of studies based on this concept is found in Rogoff, B. & Lave, J. (eds., 1988), *Everyday Cognition: Its Development in Social Context* (Cambridge, Mass., Harvard University Press; cf. also Lave, J. (1988), *Cognition in Practice: Mind, Mathematics and Culture in Everyday Life*, Cambridge, England: Cambridge University Press).

Figure 4-2: Alice in Wonderland

Turkle defines the "soft approach" to *bits* and *bytes* as *bricolage*, borrowing the term by which Lévi-Strauss habitually indicated the associative science of the concrete as practiced by numerous traditional societies[30]. But Turkle uses this concept differently. While Lèvi-Strauss proposed "bricolage technology" as a problematic borderline between the analytic method of western science and the science of the concrete that traditional societies have invented to face the power of time, Turkle on the contrary exalts its story-telling function. It is as if the computer screen evoked the phantasm of what the psychoanalyst Winnicott defined as "transitional objects", such as Linus' blanket, a threadbare rag doll, or patches of silk from a first pillow, which small children are still attached to even as they set out to explore the world beyond the crib[31]. According to Winnicott, as children grow they leave real objects behind, but the experience of these objects remains. In the course of their future life, this experience is "irradiated" into a highly "charged" intermediate space

[30] Cf. Lévi- Strauss, C. (1962), *La pensée saivage*, Plon, Paris; ed.it. (2003), *Il pensiero selvaggio,* Milano, Il Saggiatore.

[31] For Winnicott, these are in reality mediators between the child's primary tie with the mother, perceived by babies as inseparable from themselves, and the capacity to make and develop relationships with others, perceived as separate beings. Cf. Winnicott, D.W. (1971), *Playing and Reality*, New York, Basic Books; Ital.ed.,*Gioco e Realtà* (1998), Rome, Armando.

between the self and objects. Turkle transfers this experience, classically associated with religion, the sense of beauty, sexual intimacy and the sense of communication with nature, to the interface.

Internet, the web and wireless communication are not media in the traditional sense, but are instead vehicles of interactive communication. Castells (2009: 75) defines this completely new form of communication as “electronic autism”, referring to the social spaces of simulated reality that combine the sociability and desire to experiment that are typical of grassrootsconvergence with role playing games. These spaces are highly characterized by a structure in which the rapidity of the question-answer relationship constitutes the outstanding aspect of the user’s emotional involvement. Among these spaces, an absolutely unprecedented phenomenon is the worldwide spread of interactive online games. Indeed, the videogame and online interactive game industry is worth over 40 million dollars.

The online game is an intermediate area within which two fundamental variables of the cognitive frame are at play: the actor’s emotional potential, and their social connection. The sense of power, of control, and of command over the objects and situations offered by this type of game allows the user to construct a processing of primary emotions[32] that involves physiological, experiential and behavioural changes. In 2008 the largest community of online games, World of Warcraft (WOW), representing slightly more than half the MMOG (*Massive Multiplayer Online Game*) sector, and had over 10 million active members, more than half of whom lived in Asia.

Game-playing can even come into competition with real life, including direct social relations and job performance careers. The Californian game-designer Jane McGonigal (2011: 28) states that “game-playing is the exact emotional opposite of depression”. All the neurological and physiological systems that are at the foundations of equilibrium (attention systems, emotional and memory centres) are fully activated by game-playing. According to the American psychologist (1991), games make it possible to experience virtually immediately “the satisfying and exalting sensations of creative fulfilment and augmented functioning”[33]. Referring to his favourite

[32] The primary emotions are non-verbal and often unconscious. They derive from variations in the flow of energy associated with the phenomena of activation and deactivation in the mind linked to processes of evaluation. This definition of primary emotions agrees with Ross Thompson’s concept of “emotional dynamics”, which involves the immediacy, intensity and specificity of emotional responses (Thompson, 1994).

[33] Cf. the study by Csíkszentmihályi M. (1991), *Flow*: *The Psychology of Optimal*

activities—chess, basketball, climbing and ballroom dancing—he defines this as flow. Obviously, these are activities that are performed for sheer pleasure, and not for status or money or duty.

From the time when Csíkszentmihályi's study was published, two developments have taken place that should be most carefully considered: the growth of so-called "positive psychology", and the explosion of the computer and videogame industry. The game sector has created numerous scientific research laboratories exclusively dedicated to the neurobiology of emotions that are involved in game-playing; at the same time, "positive psychology" has concentrated on human resources, focusing on the optimal functioning of individuals, groups and institutions. In this context, various recent works have demonstrated that subjective well-being and "happiness" can be permanently improved[34]. Some critics of positive psychology maintain that it is too "focused on ego". It is true that numerous studies in the field of behavioural genetics suggest that 50% of subjective "happiness" is determined by psycho-physiological variables that cannot be changed. However, the remaining 50% of "happiness" can be learned, above all, through individual actions and the participation in social groups. Videogames, and online (MMORPGs – *Massively Multiplayer Online Role Playing Games*) games, make ample use of the body of knowledge that has been accumulated over the last ten years by the "happiness engineers". Game developers know very well that games are commercially successful in direct proportion to the quantity of "positive emotions" they create in their players. Since the birth of "positive psychology", the most important 'idea-persons' of the large studios take the field's research findings into consideration with respect of the development of their best games. As one journalist from *Wired*[35]has

Expe- rience, New York, Haper Perennial; cfr., inoltre, Csíkszentmihályi M. (1996), Kubey R., Larson R., M 1996), "Experience sampling Method Applications to Communication Research Questions", in *Journal of Communication*, 46 (2), pp. 99–120.

[34] In contrast to the longstanding dominant tendency in psychology, this approach does not only attempt to alleviate psychic sufferance, treat mental disturbances or remedy social pathologies. Rather, it studies the psychological mechanisms that help people grow and defend themselves from the effects of stress and psychic disturbances that can come up in the course of life. For a closer examination, see Peterson C. (2006), *Positive Therapy: a Metatheory for Positive Psychological Practice*, London & New York, Routledge; Peterson, C. (2006), *A Primer in Positive Psychology*, London & New York, Oxford University Press.

[35] Thompson, C. "Halo 3: ""How Microsoft Labs Invented a New Science of Play", in *Wired*, 21 August, 2007,

commented, "Microsoft's game-testing lab looks more like a psychology research institute than a games studio".

In 1983, the jazz pianist and sociologist Sudnow published an experimental account regarding one of the first home videogames, *Breakout,* which was a game similar in design to ping–pong[36]. In this account, he showed what was so new and spell-bounding in digital games from the emotive point of view, even before the arrival of spectacular graphics, epic stories and massive multi-player worlds. Sudnow described the extreme neuro-chemical activation that takes place in our brains and bodies when we start playing a good digital game. This is what researchers of emotion define as an arousal state (cf. chap. 2, p. 86). In his analysis, videogames took on the traditional properties of activities that are potentially able to induce this state (a goal, obstacles, a growing challenge, willing participation) and then used a combination of direct physical input(the joystick), flexible regulation of difficulty (informatics algorithms), and instantaneous visual feedback (video-graphics), to drastically shorten the chain of retroaction. This feedback chain produces a much faster cycle of learning and gratification, leading to a perfect and powerful sense of control over the "micro-world" on the screen. Here is the brief description that Sudnow (1983:20) gave of the arousal state of the videogame player:

> …the joystick-button is felt to be a genuine instrument of action. *Bam, bam, bam*, I got you … Please, don't miss him, come on, get that brick, go slow now, no surprises, calm down, don't panic, get the rhythm, now go. Take that ending. Video game action. You know when you're there. Like the first time you got drunk.

The experience of heroic tension embodied in the simulated scenarios of onlineand videogames is evidence of a progressive exodus towards the game space of hundreds of thousands of "active players", who "abandon reality" for longer and longer periods of time. This can be explained by the special nature of their dynamic-interactive structure, in which the speed of the question-answer relation is the most outstanding aspect. In action videogames,[37]the best player is the one able to react in a split-second to a request for action, while the worst is the one who cannot do this.

---

http:/www.wired.com/gaming/virtualworlds/magazine/15–09/ff-halo.

[36] Sudnow, D. (1983), *Pilgrim in the Microworld*, New York,Warner Books. The complete text is available at http://www.sudnow.com /PMW.pdf.

[37] Action videogames, like the well-known *Beat'em up*, *Shoot em up* and *First*

The challenge of digital games is to nourish the economy of mobile collaboration between the real world and the local environment. For example, Groundcrew, the wish-fulfilment project developed by Joe Edelman, a social entrepreneur from Cambridge, Massachusetts, has the job of keeping the simulated individuals happy and healthy.[38] Their health is preserved by attending to their physical needs: feeding them, putting them to sleep, and making sure they have a shower and use the toilet regularly. The logistical part is the key, as Edelman immediately realized when he founded his software company, Citizen Logistics. Taking his inspiration from *The Sims*, the most successful of all simulation games, he dedicated himself to creating a system that would make real-world wishes correspond to a simulated local environment. *Groundcrew* is the simulated space that potentially incorporates an economy based on the exchange of three intrinsic gratifications: the happiness that comes from "doing good deeds"; the thrill that accompanies the achievement of a challenging mission; and the satisfaction of accumulating points which count up the fulfilment of wishes.

Around the globe, the community of onlineplayers(counting those who play on consoles, PCs and mobile phones) numbers more than 4 million people in the Middle East, 10 million in Russia, 105 in India, 19 in Vietnam, 10 in Mexico, 13 in Central and South America, 15 in Australia, 17 in South Korea, 100 in Europe and 200 in China[39]. According to this community of players, everyday reality fails to offer the powerful social ties met in MMORPG environments. The phenomenon of online games is expanding at a dizzying rate worldwide, and it constitutes an analytic frame that can no longer be ignored by scholars of the body and of social action. A first attempt came from the English University of Nottingham Trent, specifically from the sociologist Monica Whitty and the psychologist Garry Young.[40]

Person Shooter, are distinguished from passive videogames, like chess.

[38] Monica T. Whitty, Garry Young, Lewis Goodings, "What I won't do in pixels: Examining the limits of taboo violation in MMORPGs", Division of Psychology, Nottingham Trent University, Nottingham NG1 5LT, UK in *Computers in Human Behavior* 27 (2011) 268–275 - *Computers in Human Behavior Journal* homepage: www.elsevier.com/locate/comphumbeh.

[39] To consult papers regarding this sector and market research studies done in the last three years, cf. McGonigal, J., (2011), *Reality is Broken Why Games Make Us Better and How They Can Change the World*, New York, The Penguin Press.

[40] Cf. Garry Young , Monica T. Whitty "Games without Frontiers: On the Moral and Psychological Implications of Violating Taboos within Multiplayer Virtual Spaces", Division of Psychology, Nottingham Trent University, UK, in Computers in Human Behavior", 27 (2011) 268–275 journal homepage: www.elsevier.com/locate/comphumbeh.

Do online games offer liberation from social constraints? Or do they confirm and reinforce the social stereotypes which we only delude ourselves that we escape from? In online games, can we violate the taboos that set constraints in the real world?

These are the questions addressed by the MMORPG's research methodology in the last five years, by these authors, starting from the supposition that "the fundamental essence of play is the license to create something *beyond* so-called normal reality" (Whitty, Young, Goodens, "What I Want Do in Pixels: Examining the Limits of Taboo Violation in MMORPGs", *Computers in Human Behavior*, 2011, 27: 269). Where a relation is contingent, there exists the possibility for change—for things to be other than they are. This is the scope of gamespace. Thus, if you do not like a particular aspect of the virtual world you are in, then you can either change the contingency relation to suit what you do like, or find another space in which this contingency already holds. One advantage of gamespace over our offline world, then, is that a change to a contingency relation is more immediately obtainable—what we call *virtual immediacy*. Based on this assumption, the empirical focus of the quoted research is whether participants can do this, with respect of how consensual sex is portrayed, along with blasphemous behaviour and aggressively expressed social discrimination. The responses of the interviewed sample group (5 adults over 18, 4 of whom were males) show the impossibility of emotionally separating the games examined from the real world (see Table 1).

Table 4-1 Journal homepage: www.elsevier.com/locate/comphumbeh

Participant	Game	Age	Sex	Years playing	Average hours per week
1	WoW	19	M	4	26
2	Wow	32	M	4	19
3	Wow	21	M	4	20
5	Sociolotron	41	M	2	28

In later studies, such as those collected in the volume *Transcending Taboos*: *A Moral and Psychological Examination of Cyberspace* (2012), the authors extend their investigation to social webs such as Chatroulette, examining the ethical and behavioural problem that results from the impact with violent and/or illicit games. The theoretical framework that emerges in this volume can be summed up by the question, "Are we able to deal with the potential moral freedoms offered by cyberspace"?

To this question, which comes out of the fundamental conflict of modernity between "life and rules" (Rodotà, 2006), between the exceptional and the everyday, right and wrong, law and violence, we can only reply with a heartfelt invitation to the social sciences to drastically revise their way of "thinking about technology" as it relates to the constitution of our deepest feeling. Nowadays there is an increasingly widespread idea that technology is an exogenous fact, a factor influencing the "world we live in" from outside – something outside of us, but also and above all something over us. A rather obvious fact tends to be ignored: technology is not, as it were, in the wilderness, outside of society; rather, it is located inside society and is, fortunately, conditioned by its social, economic and cultural dynamics. We need a sensorial re-mappingof our relations with computer technologies in order to create a new cultural paradigm that will allow us to "domesticate"[41]the emotional spaces revealed by online game practices. At the same time, we need to keep well in mind the suggestion of Antonio Damasio (2003: 206), according to whom "the mind of the brain, furnished by the body, governed by the body, is the servant of the whole body".

4.3 The Asynchronous Maps of the Mobile Interface

The term "mobile" has been applied to technologies since the age of the papyrus, when the written word first became transportable far and wide. Today we tend to apply this word to digital devices such as mobile phones, GPS units, tablets, Ipads and game-systems. The notion that mobile technologies are new is extremely near-sighted. Throughout history, whenever a medium that was once known to be geographically fixed became mobile, a cultural change has accompanied this technological change. From the moment the written word was moved from stone inscriptions to papyrus, the world changed. The introduction of writing did not limit itself to moving the sensory focus[42]. Besides moving cognitive

---

[41] The concept of "domestication" is employed by the English anthropologist James Goody, as well as by Leroi-Gourhan, *Il gesto e la parola*, quoted in chap.1), to indicate the work a culture performs in organizing time. Goody specifically connects the concept to the changes in mental and social structures brought about by the use of writing (cf. Goody, J., 1981, *The Domestication of the Savage Mind*, Cambridge, Cambridge University Press; Ital. ed., *L'addomesticamento del pensiero selvaggio*, Milan, Angeli.

[42] In oral-auditory culture, the word is an active force connected to the present. On the contrary, in alphabetic-visual culture, the word is transformed into a *mental meaning* without temporal constraints. According to Ong, an American sociologist

processes, it also had important effects on social structure, from the Roman Empire to the development of ecclesiastical power. The passage from primary orality to the chirograph culture is, above all, characterized by a sensory shift (Ong, 1986) from hearing to sight. If oral culture plunges the social actor into an immediate world of sounds, writing places an observer *outside* of what they see, at a distance. To be exact, there is a transition from speaking, which is an action, to writing, which becomes memory and reflection (Goody, 1986).

This analysis confirms the notion of brainframe, by which de Kerckove (1993: 11) defines the *spatial frame* of the alphabetic consciousness. For this Canadian scholar, the invention of Renaissance perspective constitutes "the reversed mirror image of the organization of alphabetic consciousness" (de Kerckhove, *ibid*: 41). Thanks to its sequential properties, the alphabetic brainframe"carves out the world" in the colours of history, geography, grammar, philosophy, art, architecture, and virtually all the other branches of knowledge. In fact, starting from the Greek alphabet, Western culture is the only one that has to some degree concerned itself with representing the world in perspective. Chinese culture, along with the cultures of Egypt and the rest of Africa, has been almost completely indifferent to perspective. Nor did it occur to our culture during the Middle Ages. The taste for perspective was only discovered in periods when literacy took a leap forward, as when reading the alphabet was taught during the golden age of ancient Greece, and once again around the time of the invention of printing. The discoveries of perspective, first at the turn of the 6th and 5th centuries B.C. in Greece, and again from the end of the 12th century through the late Renaissance, can indeed afford the best example of how the alphabet has "restructured" our minds[43].

The notion of brainframe presupposes a conception of technology as an *extension* or *amplification* of the sensomotorary, psychological and cognitive functions of the mind. For example, for de Kerckhove, television is an extension of the eye and the ear to the places the images come from. That is, it is a "collective imagination projected outside of people's bodies (1993:23) so that it comes together in a relatively convergent process. The

---

of the Toronto school (1986: 136), the sequential and spatial analysis of the word that started with writing and "raised to a new order of intensity by printing" is further amplified by the computer, which maximizes the ties of the word to space and local (electronic) movement.

[43] In a conversation with Jonathan Miller, Ernst Gombrich suggested that the refining of perspective representation grew out of a need for realism, which he called "the principle of ocular testimony": the wish to represent and see an event exactly as if it were happening before one's eyes (Miller, 1983: 231).

methodology proposed by de Kerckhove considers culture as absorption of the so-called "technological extensions" which, by involving language and the way we use it, also influence our strategies of information processing. De Kerckhove in fact maintains that only by adding sensory interactions can we reconstitute, *outside* our body, the "type of interiority" that is characteristic of human consciousness.

The theory of the alphabetic brain frame presupposes the notion that the technological instrument is exterior to the body's sensoriality. Its underlying idea is that information elaboration technologies "frame" our brain in a structure in which the dialogue with culture, both physiologically at the level of neuronal organization, and psychologically at the level of cognitive organization, is condensed. According to De Kerkhove, a main feature of the cybernetic brainframeis the exteriorization of consciousness. In particular, the possibility of using the Internet to access information present at any point in the environment leads to the development of a "connective intelligence", in which the production of knowledge and culture is socialized. The "subject" identifies new possibilities, and solves problems using the multiplying power of the social network, thereby sharing the thoughts, intentions and projects expressed by others. In this context, De Kerkhove speaks of "technopsychology", which is meant to clarify the ways in which the human mind demonstrates its dependence on dominant technologies. He asserts that against this backdrop, aesthetic experience takes on a precise function, in that art has an important role as practice, in the reconfiguration of our brainframes. In the electronic era, the transition from the linear model imposed by writing and reinforced by printing to become a new connective and reticular brainframe, involves significant modifications in our way of organizing experience[44]. The risk that the transition from the discreet and linear to the connective and reticular model may, in the end, reduce our capacity to give a syntactic order, an experience that is available to one and all. In this connection, we need only think of the difficulty that younger generations have with writing and, above all, with paying attention.

Tracing the boundary between the social sensorium[45]and contemporary technologies is, therefore, the methodological problem that comes out of

---

[44] In brief, on this view the emergence of today's brainframeshas challenged the linearity of time, emphasizing the primacy of real time and thus producing two effects: 1. on the one hand, brainframes favour the acceleration and fragmentation of temporality; 2. on the other hand, they shift the focus onto the present, leaving past and future in the background.

[45] In English, the term sensoriumis used to indicate the complex of sensory

the emerging spatial turn, to use the words of Warf, Arias (2009). He speaks of the "spatial turn" that is developing through the reorientation of the classical digital paradigm (a concept that is, by contrast, centred around the idea of the disappearance of space, the end of geography, the reduction of mobility, and the annulment of incorporation in digital network communications)[46]. Starting from the 1970s, this "boundary" has been a "bone of contention" between the so-called psycho-technologists (or the Toronto School), and social constructivism[47].

---

functions connecting the human body with the context in which it moves and acts in a given moment (cf. *Dizionario il Ragazzini* ,Ragazzini, G., 2005, Bologna, Zanichelli, p. 972).

[46] Two contrasting visions of the relationship between space and digital network realities have met and locked horns in the last twenty years (1991-2011), both in research labs and in market competition. The first of these has aimed to take the world into the computer, simulating virtual realities and synthetic environments online, whereas the second has tried to take the computer into the world by incorporating calculating abilities into physical architectures and concrete objects. Mark Weiser, the father of ubiquitous computing, has defined the former as virtual reality and the second as embodied reality.

According to Weiser, the authentic destiny of a way of computingthat is useful to man is to recede into the background of the natural environments he inhabits. The so-call "third wave" of computing was started by the PARC group, starting from the term ubiquitous: every person served by many computers, and not one person for every computer (the PC by now in notable decline). This vision implies the possibility of accessing necessary information and high-capacity elaboration tools (remote, local or portable/wearable) at any time, in any place and in any situation, through a large variety of media and devices (Dourish, 2001). This translates into the integration of computer processing capacity in the most common objects (such as pens, to memorize and process what we write in digital form; clothing labels, to check their position and washing instructions; and light switches, to be turned on and off at a distance.).

On further consideration, we see that these two paradigms are based on two distinct ideas of space and virtuality. For the promoters of virtual reality, real space is compressed and cancelled by network information technologies and real time. In this framework, geography, distance, mobility and corporeity lose their value and are substituted by interactions and virtual worlds. The latter are three-dimensional simulated worlds seen on computer screens or through special viewers that immerse us in simulated environments. This is a dimension in which real time and virtual space coincide, nourishing each other reciprocally. For if real time is the instantaneity of communication and action at a distance that is the outstanding characteristic of our day, virtual space is the immaterial geography that constitutes the space of data.

[47] According to the first group of scholars, media are a type of technology, an "extension of ourselves". According to the social constructivism approach, whose

In the evolution of contemporary mobile communication, the concepts of space/code, integrated by the communication and social revolution of the amplification and portability of social networks, represents the foundation of the new paradigms of mobility[48] which are being produced in Post-PC society through technological evolution, and more specifically, the ecologies of wirelesscommunication. The following factors contribute to this "turn":

1. the strengthening of mobile telecommunication technologies and of access to the Internet from mobile devices; the spread of mobile devices(smartphones and tablets)that are increasingly sophisticated, so much so that they create an aesthetic-experiential relationship between the consumer and the objects of consumption and communication; the

---

main reference is Williams (19081), the critical element that characterizes a medium is *how and why* it is used. Like Mc Luhan, Williams identifies the essence of media in their nature as technology. But while for Mc Luhan and his followers the main characteristic of the medium is its *physicality*, for the constraints it imposes on perception and interaction, Williams shifts the focusonto the utility of technology within a given context. According to McLuhan (1962), the genesis of a medium has a psychobiological origin: people react to stress by "numbing" the part of the body that is under stress. Immediately after, they try to produce a prosthesis – a technology or a medium – to externalize the component of the body that is stressed. On the contrary, Williams feels that the medium has a psychosocial origin: it is born of the development and reconfiguration of the resources of a culture in order to reach a socially defined goal.

[48] The impact of physical mobility on the logic of contemporary cultural is such that sociological research is reorienting its analytical paradigms in order make mobility the founding paradigm of our everyday reality. In this new perspective, the movement of people (commuting, travelling for work and for for tourism, migration, flash mobs*)* and things (transfers of goods and capital, means of transport, networks and circulation of contents and services) bring together the "paradigm of new mobilities", both material and communicative. By "paradigm of new mobilities" is meant an analytic approach that shifts attention from the traditional sedentary and static nature of social structures to the flow, movement, circulation and dislocation of things and persons. Ever more discussion is dedicated to mobile societies, mobile lifestyles, mobile media and devices. In our day, these are mobilities that produce the meeting of places and people legitimating and strengthening relations between consumers and brands and mediating our sense of proximity to everyday objects and events. By doing so, the new mobilities also re-invent the practices of co-proximity and socialization of consumers and users, including the form of social networks activated with mobile devices (practices technically defined as mobile social networking or locative social media).

spread of contents and services in the form of applications, which shift the utility, personalization and contextualization of experiences with the extreme usability of interfaces; the portability of social graphs (Hoffman, Novak, 2011), meaning the possibility of bringing the social networks' power of connection into the places and experiences of single interactions;

2. the appearance (or reappearance with innovative features) of services and applications on a geo-locative basis (Location Based Services, LSB); these are platforms activated by increasingly sophisticated cartographic and geo-information systems, recently also employed to socialize personal geographical positions (Salt, 2011a);

3. the progressive implementation of Ambient Intelligence, AI; this means computer architecture incorporated in everyday objects thanks to context-aware technologies, which create growing spaces of information (Augmented Reality, AR).

To further examine each of these factors:

1. Nowadays the Internet is a three-tier ecosystem. The first tier is made up of a central core of high-capacity, relatively stable interconnected computer networks. A second group is formed by more local networks, and a growing number of mobile connections that are added on to the main core. The third tier consists of all the intelligent devices and accessories that are connected to local and mobile networks. This tier is what in jargon is known as the Internet of Things, made up of the exponentially growing number of objects and goods (cars, household appliances, accessories) that are connected to the web in order to receive feedback, process data, and communicate information. Recent estimates indicate that there will be fifteen billion intelligent objects connected to Internet in 2015[49]. The publication and social sharing of the position of consumers (social disclosure of spatial information) is the lever that companies like Pepsi, Virgin, the History Channel, the New York Times, Nike and many others propose to exploit for their business goals. *Hybrid spaces* is the term Adriana de Souza and Silva (2009) have used for these mobile realities, created by the dislocation of users with mobile devices, who are continually connected to Internet and to the social networks they belong to. These authors use this concept to stress the specificity of the mobilesocial spaces created by instruments like smartphones, tablets and game

[49]This estimate is found in a recent report of IDC (2010).

consoles. What is pertinent here is that thanks to wireless connection and services, this new space is created and updated instantly, and enriched by the mobility and sociality of the users, which is mediated by mobile devices and interfaces;

2. The mobile and wireless world, like all present and future handheldor wearabledevices, is the heart of the locative media, a topic of much interest and research on the part of the international scientific community[50].Though difficult to define precisely, locative media can be seen from the technological standpoint as all of the appliances, applications and services that have resulted from the merging of pervasive, ubiquitous and wearable computing with "geo-localization",[51] a function made possible by mobile positioning and tracing technologies. Why has this merging come about? Because it permits the association of two fundamental sources of knowledge, namely "information" and "place in geographical space", thereby achieving the "territorialisation of information". This highly powerful combination produces a very large number of applications, as well as a totally new way of understanding surrounding reality. For instance, Google Earth replies to a place search with real images, along with information attached to the images. The universe of locative mediais increasingly full of events (projects, centres of competence) and is in continual expansion. It goes from the core of "lighthouse-projects"[52]towards geo-localisation services and the new frontiers of Ambient Intelligence;

3. The reservoir of developments representing the various fields of activity of Ambient Intelligence[53](AmI from here on) constitutes the

---

[50] Some important institutions for reference include:

- •RXC (Riga Center for New Media Culture (http://rixc.lv), the Latvian competence centre, founder of this field of development;
- Locative.net, the collection point for all these studies [http://locative.net];
- •The international conferences of ISEA *(InterSociety for Electronic Arts*) 2004 and 2006 (http://isea2006.sjsu.edu) and *PerCom* 2006;
- *IEEE International Conference on Pervasive Computing and Communication* (in Italy - lia, http:www.percom.org/).

[51] Geo-localization is the mapping by a technical system of the position of an entity on the earth's surface, in terms of its geographical coordinates. GPS (*Global Positioning System*), by which car navigators work, is one system belonging to this domain.

[52] The following are references indicative of the potentiality of locative media: *Urban Tapestries*

[53] Ambient Intelligenceis a concept in continual evolution, thanks to technological innovation on the one hand and changes of a socio-economic nature on the other. AmI is the universally used abbreviation for Ambient Intelligence, where the "m"

most significant scenario for applications. With the aim of improving lives through the use of advanced technological infrastructures, AmI systems offer a kind of service called "butlerservices"[54]. These are meant to facilitate users in the performance of everyday actions, as well asmaking new ones possible by adapting to their personality and the activities they perform.[55] From the perspective of AmI, knowledge means learning, awareness of the context of action, and access to information. It is used in a way that is appropriate to the context and to an individual's needs. Speaking about knowledge in conditions of mobility, the following aspects emerge:

a) the sense of mobility of the "fount" of knowledge (human experts);
b) the variability of the place where a learning activity happens;
c) the change of place where knowledge-intensiveprojects are carried out.

AmI technologies set themselves the goal of supporting individuals – who are both theproducers and the consumers of knowledge – in the processes of the production of, and search for, information, helping them to manage the unexpected and variable situations connected with mobility by *concealing the contextual aspects that require only routine management* in order to highlight others that can foster processes of knowledge. The vastness and overlapping of AmI scenarios make it difficult to classify them. Interpretation in terms of fields or dominions or "thematic areas" is not homogeneous, as it depends on the objectives of the organizations involved and the quantity of research funds available. The potential contributions of AmI to our way of life are clearly related to knowledge, and more precisely, to the experience of knowledge. Its impact on knowledge is largely due to the distinctive features of AmI digital applications, i.e. multisensorialityand multimodality.[56] These features are

---

is included to distinguish it from AI, the acronym for Artificial Intelligence.

[54] Cf. project GENIO, chap. 3, paragraph "Casa", Hartmann, 2004.

[55] In brief, the sectors of analysis are as follows: knowledge management, work, tourism, home, free time, media and emergency situations.

[56] Multimodality, multisensoriality (and multimediality) are correlated terms, but they are not synonyms. By "multimodality" is meant the many modes that can insert input into a computing device (typing in, mouse, writing, voice, gesture, expression). A determined mode can be conveyed by different sensory channels (multisensoriality), among which the most common, but not the only, ones are voice, hearing and sight. The widely used term "multimediality" usually refers to

essential, in that they amplify, multiply and combine the channels through which input/output between individuals and everything external to them takes place, thereby fostering the acquisition and metabolisation of knowledge.

The so-called "spaces of knowledge" are the expression of these modalities. Abandoning the conventional arrangement of the by-now extremely common screen - keyboard interfaces, AmIs attempt different approaches to define the interface. A particular definition of knowledge widely used among computer experts is that "knowledge is information in context" (Nyiri, 2002). We can take as an example the European Mummy project, which is an important model in this sector[57].

The project's goal is to improve the efficiency of the business processes in which the mobility of people and objects is important, with the object of facilitating the acquisition and transfer of knowledge. The scenario referred to is the activity of professionals with the profileof knowledge workers, like architects and engineers, whose jobs require moving about frequently. In the presence of this type of situation, common to various categories of workers, the project proposes a system of Knowledge Management(KM) provided with a series of functions, accessible to the user by palm-top, which integrates multimedial technologies that have systems of annotation and collaboration, technologies of Context Awarenessand technologies of Knowledge Management.

The re-emergence of spaciality, thanks to the above factors of technological innovation in various disciplinary domains (geography, sociology, architecture, economics, art and urban informatics, to name some), does not, however, imply a return to the concept of space understood as a mere container of individuals, objects and events. This return of the relevance of spatiality, of physical place and of the offline world, is not a return to space as we have understood it up to now. The "new" space brings with it all the potency of information and communication produced by digitality. In fact, the old opposition between one extreme of virtuality, understood as simulation of the real world, and the other of materiality, understood as static geometric space, is less and

---

the digital channel that conveys information (for example, an image on a screen, web animation, video streaming, sound over the air).

[57] The objective of the Mummy (*Mobile Knowledge Management*) project is the improvement of the efficiency of business processes, implying the mobility of persons, through the proposal of mobile knowledge managementsystems based on the use of multimedial technologies. Cf. Mummy, *Mobile Knowledge Management*, Homepage, 2005, www.mummy–project.org/ Partner: University of Prague (Cz), University of Waedenswil (CH), COSMOTE(Gr), Arcadis (De).

less able to capture the new realities, which are fixed and digital at the same time, as citizens and consumers are now experiencing. In actual fact, these two extremes of virtual and real do not exist today. Rather, there is a *continuum* of mixed realities (MR). This term indicates the environments and practices that in various ways join spatial reality and digital information. In particular, the idea emerges that space, being relational, is to be considered as "transducted space", i.e., space continually and socially constructed, and activated by the networks and new digital realities emerging at the crossroads of mobility, geosociality and "augmented realty"[58].

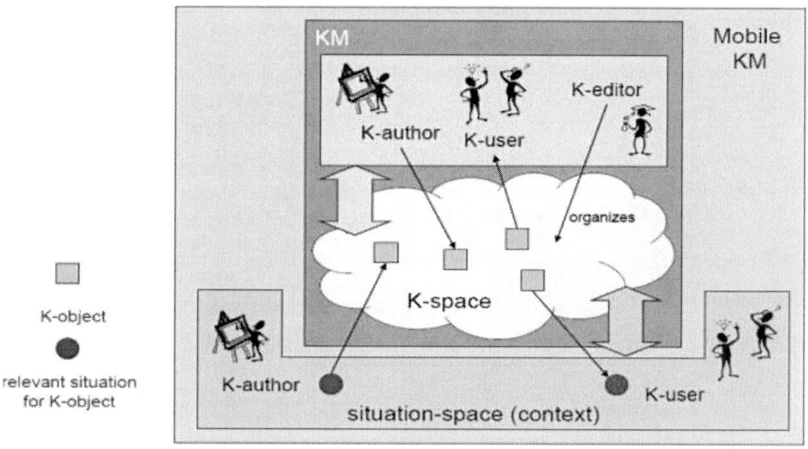

Figure 4-3: *White Paper: Mobile Knowledge Management, 2004, Mummy Project*, Workshop SIe-L-Florence, 30.03.07. Ambient Intelligence: tecnologie e realizzazioni–Federica Paganelli 10.

[58] A group of researchers and developers, working together in the Acceleration Studies Foundation (http://www.accelerating.org/) on the *Roadmap* project has classified four types of metaverses:

1. "augmented realty", in that it is structured as "superimposition of a new, virtual level of information on the real, the physical, material" (Granieri, 2009: 73);
2. so-called life-logging, i.e. the metaverse in which our personal and vital reality is transposed into *networks*, employing, for example, blogs or social networks;
3. so-called "mirror-worlds": the type of'metaverse referred to instruments like Google Earth, which re-propose reality in the form of information, allowing us to navigate it;
4. virtual worlds, which simulate and augment reality as we know it.

More and more, mobileInternet is becoming a network of the social mapsthat animate our spaces, as platforms of location-sharing,thanks to which users choose to share with others (friends, contacts, family members, brands, marketers) their metropolitan movements, preferences for events offered in their territory, and opportunities for encounters. Space is no longer simply a static geometrical expression. Rather, it is modulated through the creative activity of the code that has been projected and grafted into objects and processes. To describe this process, which generates new specialities via software programming, advanced digital theory speaks of transduction of space[59].

The key point is how space is produced today and, obviously, with what new configurations it is experienced by users and the general public, thanks to the software code that has been incorporated. The current phase of transformation for wireless networks, platforms of locative services, mobility devices, and socialized and augmented practices, creates a forecast of a near future that is particularly complex and open to challenges that may be risky (at least as regards political and economic democracy), but that have great potential.

Analysis of the new mobile socio-medialities requires up-to-date approaches and research models that have been designed for the new contexts and practices. The risk lies in wanting to investigate and manage the new phenomena with cognitive and methodological tools that are not aligned to the analytical challenges of our day, and a tomorrow that is not so far off. We are undoubtedly less careful than we were fifteen years ago about the spaces of freedom and social imagination that the web has brought into our lives, from the time when the page network started to face the competition of the apps that mobile technologies have popularized. But it is not enough to preserve net neutrality(equal access to the net for all operators and contents). We must also monitor net generativity, i.e., the opportunity for users and developers to use the creative and agentive capacity characteristic of digital technologies and tools, while balancing freedom of choice with efficiency.

Within the panorama designed up to now, the body is the *multisensory integrator* between sensomotory activity, and the simultaneous connection produced by a specific mobile technology. The fundamental feature of this simultaneous connection is the *asynchrony* between the time experienced by actorsof a single technological performanceand the requirements of real

[59] The theory of *transduction of space* is taken up anew in Dodge, Kittchin (2011), an essay in which the authors refer to their previous essays and to the work of A. Mackenzie (2002, 2006) on the social dimensions of software and programming codes.

time. For example, in performanceslike Rider Spoke[60], the body appears as a media *artefact*, which means that it is presented as the result of an action of mediatisation rather than as the condition of priority which culture has to enact in order to achieve the *sociosomatic balance* of the human being in the various contexts "in movement" today.

Figure 4-4: www.blasttheory.co.uk/.../work_rider_spoke.ht

This body as "artefact" serves as the compass of the new alliance between human senses and mobile media, between information and entertainment, work and play, within the current "entertainment regime". The origin of this alliance is the representational screen that governs the theory of "embodied cognition"[61], which can be traced back to the phenomenology of Maurice Merleau–Ponty and, with the digitalization of the world, defined by the so-called relational media. This theory posits that what "I" perceive in the *hic* and *nunc,*depends on what I am doing. From one moment to the next, the world offers me "affordances"[62], opportunities "translated" by the man-

[60] Rider Spoke is a mixed reality performance developed by Blast Theory in collaboration with the Mixed Reality Lab. It was put on stage for the first time at the Barbican in London (October 2007) as a creation for cyclists. It combines elements of theatre, performance, game playand state-of-art technology. Rider Spoke has since then been presented in Athens (2008), Brighton (2008), Budapest (2008), Sydney (2009, Adelaide (2009) and Liverpool (2010).

[61] This theory considers corporality as a necessary condition for the development of cognitive and social processes, meaning by the latter all of the organism's sensory-motor capacities that allow successful integration with an environment (cf. Lakoff and Johnson, 1999; Niedenthal *et al.*, 2005; Morganti, Riva, 2006).

[62] The term affordancewas first introduced by the cognitive psychologist Gibson (1979): an affordanceis a resource that the environment "offers" to a "subject" who

computer-culture interface[63](Manovich, 2001: 98). If, in the past, the great mechanism of the universe was the brand, perhaps in our day it is the silent and largely invisible network of networks—Internet and the entire mobile infrastructural network it is based on. Time furnishes an organizing frame and structures associations. As every traveller knows, the sense of time and the propriety of behaviour change drastically from one culture to another, and among various sub-cultures in pluralistic societies.

Being in synchrony with time requires us to focus on the rhythms, events and celebrations that connect a community with the rhythms of work and play.

Historically, our experience of time changed with the introduction of instruments of measurement—from the tolling of church bells, to the hands of the clock in the main town square, to the ubiquity of the wristwatch. Starting from the sundial to the hourglass, we have arrived at ultra-technological chronometers, at micro-measurement; and since speed is prestigious, from the case of sports records calculated in hundredths of seconds to that of neutrons. But this acceleration of social time frightens us, because we know that, despite our technological instruments, something gets by us—the occurrence of the contingent, and the weight that the past takes on in the domain of its changeability. Considering this disappearance of a sense of continuity and historical memory, the only role left for the historian is, as Foucault was wont to repeat, that of the "archaeologist of the past" who uncovers its traces, as does Borges in his fiction, and places them, all in a row, in the "museum" of modern knowledge. Herein lies the urgency of the question which, according to Foucault, authorizes us to catalogue most of modern philosophy as "ontology of the present", that is to say, *what is happening now?*

This "now" throws open the luminescent universe of asynchronous maps which, according to Jason Farman (2011), the sensory inscribed bodystages as "geo-socialized" emotional and aesthetic experiences.

---

is able to take advantage of it. According to Gibson, every object and environment is characterized by a series of properties that support a particular type of action and not others. Affordancewas initially considered exclusively on an adaptive basis, as the fruit of exploration whereby through its activity the body gets the information it needs directly from the environment. Later, the concept was broadened to include the process of interpretation, mediated by meaning. This is called mediated affordance.

[63] According to Bonsiepe's by-now classic definition (1995: 48–49), "the interface isthe space in which the interaction between human body, utensil (object artefacts and communicative artefacts) and purpose of an action is organized".

Within this iridescent universe, real time becomes "the last star" (Lovink 2011: 15). As if to say: time accelerates and we abandon history.

4.4 The Sensory Inscribed Body: "technology is anything that wasn't around when you were born"

In the conclusion of the volume, *Mobile Interface Theory: Embodied Space and Locative Media,* Farman quotes the maxim formulated by the computer scientistAlan Kays: "technology is anything that wasn't around when you were born" (Kays 1996: 33). In the last two decades, communication and media experts have tried to confirm this maxim, sparking an important (though not new) discussion about the ways in which the media are changing how we see the environment, our fellowmen and ourselves. Assuming the main focus to be the growing spread of mobile technologies analyzed in the previous section – including, above all, the mobile GPS[64]– the above-mentioned volume advances the hypothesis that locative media have changed our experience of space. The concept of the sensory inscribed body, which Farman proposes to analyze this change, can be summed up in three points:

> 1. Embodiment[65]and space are co-constitutive. It is impossible to inhabit space, because this would imply entering a pre-existent field, which our bodies simply feel. Rather, space is said to be constructed simultaneously with our sense of "embodiment";
> 2. Embodiment is relational: we construct our sense of space by attuning ourselves to different parts of the world, just as we perceive our bodies by selecting sensations or specific areas, rather than by "storing everything inside";
> 3. Mobile technologies reconfigure the ways users"embody" space, specifically when they allow the Internet to move around in everyday environments. Within this "embodied" environment, "subjects" situate themselves simultaneously in digital space and in material space, with one "sculpting" the perceptions of the other.

According to Farman, "we read" our bodies selectively, and never completely. Ideally, the technology in use becomes invisible in these

---

[64] The Global Positioning System (GPS) uses satellites to establish our position rapidly and within a range of only a few metres.

[65] The theory of embodiment has as its main objective the use of corporeity to facilitate the man-computer interaction (Dourish, 2001).

interactions: if the technology is working correctly, we interact with it *invisibly*. Like space itself, the medium becomes a "natural" part of the field in which the body moves. In practice, the map[66]becomes both the means and the product of the new space that mobile technologies are introducing. From location-awaretechnologies to a renewed interest in proximity in online social interactions, the map determines the specificity of a place and the cultural capital connected with it. With maps, the power of the specificity of place has generated growing interest in looking for ways to transform the scenarios that surround us into informative interfaces. In other words, accessing information on Wikipedia, from one's desk, is an entirely different experience from accessing exactly the same information with a site-specificinteraction using a mobile device.

Recently, this concept has been literally interpreted with the creation of the N Building in Tokyo. The façade of this 24-storey building is a large QR (Quick Response) code, which allows mobile devices to carry out the scansion of the façade, and obtain detailed information about what is happening inside in real time. The QR codes on the building, which can hold larger and more various characters than a standard UPC bar-code, can connect up with URLs and AR information. When people scan the N Building's façade, they can connect by twitter feeds to people inside the building, localized with 'geotag'.The shops inside the building also offer real-time information on an AR interface. This is a significant sample of an urban scenario that is at the same time an informative interface.

Farman also holds that our experience of places through mobile technologies constitutes a phenomenological involvement in the specific means of communication, at the same time as it is a way of reading the meaning of this form of involvement. Our bodies perceive the world as collaboration between material and digital space, while simultaneously interacting with the cultural data inscribed in experience, suchas the implications of privacy associated with revealing one's position to businesses and other users. The maps for mobile devices are culturally localized, and different maps of the same place give a very different sense

[66] From printed representations of a place to words scribbled on a piece of paper to note down reference points, to GPS navigation systems, our travelling through space has long been interpreted as the correspondence between the material world and the ways we represent the world. We trust maps to offer us an external visualization of our internal sense of proprioception. Maps indiate where we are in relation to the places that surround us. As Paul Jahshan notes in his book, *Cybermapping and the Writing of Myth,* though cyberspace is a place where people live, we have to struggle with orientation, and so "the problem of maps is inevitable".

of localization[67]. In these maps the user is involved in a wide variety of roles, ranging from "the disembodied *voyeur*" to the "localized subject". The mobile interface offers a decidedly *fluid* relation between the space of experience and the representation of space.

In this regard, a significant instance is Christian Nold's *Biomapping* project, which is a perfect example of the representation of space operating in collaboration with mobile technologies and visualized information. Begun in 2004, this project has so far been carried out in four cities: San Francisco, Paris, Greenwich and Stockport. The participants wear a mobile device employing a technology similar to that of a copygraph, which traces their skin response (GSR) while at the same time registering their GPS coordinates, and recording their positions as they walk around the city. As they walk, their emotional level is visualised on a map where every person is represented by a sort of differently coloured wall on which these movements are traced, with peaks where the person feels intense physical experiences.

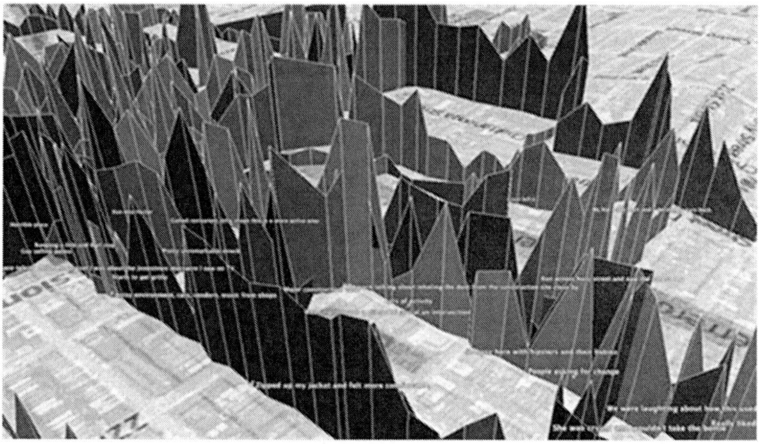

Figure 4-5: From Christian Nold's San Francisco Biomapping project: visualization of the emotionsfelt by pedestrians walking around the city

[67] A famous example of this type of relationship between representation of space and conception of space can be seen in the difference between Mercator's 1568 map of the globe and the map in the Gall-Peters' 1973 projection. In the Mercator projection, used for ship navigation, the surface of some continents, like Africa and South America, seem smaller than that of territories closer to the poles, like Greenland. Gall and Peters attempted to remedy this discrepancy by creating a map that would represent the surface of every continent more faithfully.

In an article on the use of informative visualisation in art, Tom Corby makes a good summary of the procedures underlying *Biomapping.* During a workshop held at the University of Westminster, student volunteers were introduced to the project. Each one was provided with a device and sent out for an hour to walk around the university and its nearby surroundings, which comprised a hospital and a densely populated city neighbourhood. While this was going on, each student's change of position and level of excitement was sampled every 4 seconds by the device, and the data were memorized. The information was submitted to Google Earth and visualised in three-dimensional images designed to produce what Nold calls the "emotional map" of the area. While moving around (and also after the event, while reviewing the map), participants could make notes on the positions they felt to be areas that had aroused strong emotions. In San Francisco, for example, one participant marked an area with the note, "Saw an ambulance and decided to follow it", followed by "I managed to see the person on the stretcher". As Chris Perkins observes:

> There is a relationship between the GSR response and emotive excitement: anger, alarm, fear and sexual emotions can all produce similar responses. Using this system, it is possible to construct individual paths and visualise the crests and valleys of emotion on a map. The results can be fused in composite maps that reflect wider social responses.

Corby develops the consequences of the maps and the notes generated by the users in Biomapping. These conversations have led to the accumulation of numerous plausible results. For example, consensual interpretative processes allow participants to achieve a deeper sense of ownership of their data. When the analysis and gathering of information is taken out of the hands of experts, people normally considered only as "subjects" of research studies are empowered to make and interpret results in whatever way they feel is right. In Biomapping,the sharp contrast between the aerial map and visualisation of participants' emotions, shows the equally strong juxtaposition of the map's capacity for direct reference (or indexicality)—especially if it is a map that uses aerial or satellite photographs—and the idea that the maps create changing, lived experiences. What is supplied, along with the satellite map, is a visualisation of human movement through the space of the map. As the Biomappingproject demonstrates, the map incorporates an experienced environment (for example, the fact of being a pedestrian in a crowded urban environment). This constitutes a sensory inscribed experience. My body in the city is "inscribed" in the technologies that localise it.

Another interesting art project aimed at defining the uses of geo-localised maps is "San Francisco ßà Baghdad" by Paula Levine, part of her series entitled *Shadows from Another Place* (2010). This project superimposes a map of the city of Baghdad over one of San Francisco. For every place where bombing took place in Baghdad, Levine marked the corresponding points on the coordinates of San Francisco. Once the corresponding nodes were created, a *geocaché* container was placed at every position. The overall objective of the work is to make the familiar "foreign". As Levine says:

> By collapsing the distinctions between 'foreign' and 'domestic', these maps connect the local and the global, allowing pedestrians/spectators to feel spatial and narrative contiguity between separate, distant places.

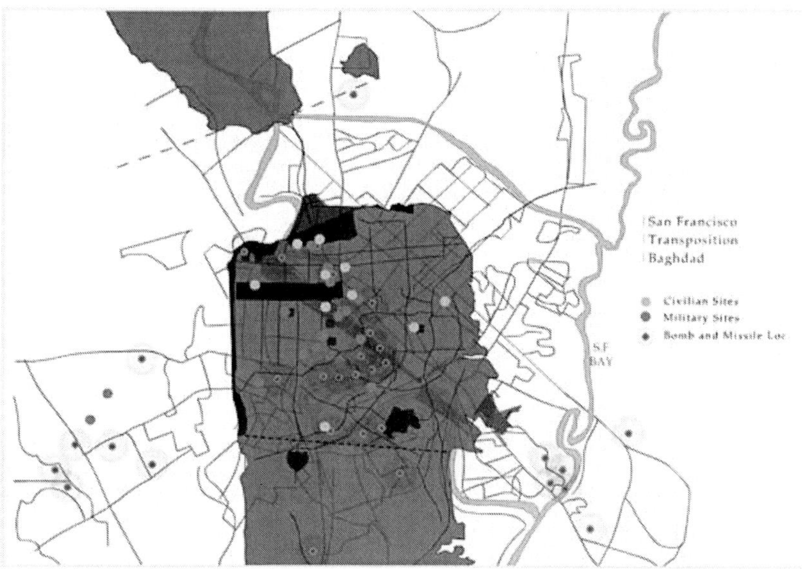

Figure4-6: From Paula Levine, *San Francisco ßà Baghdad* project (http://shadowsfromanotherplace.net).

The "sensory inscribed" maps examined by Farman are "wanderers" in the sense given by Guy Debord[68], and are totally insubordinate as regards habitual customs. As Debord said (2008: 246) back in the 1920s:

[68] Guy-Ernest Debord (1931-1994), French writer, director and philosopher, was among the founders of the Letterist International and the Situaltionist International.

> Recently a friend told me that he had just wandered around the regions of the Harz Mountains in Germany, blindly following the indications of a map of London. This sort of game is obviously only a small start compared to the total creation of architecture and urbanization that one day will be in everyone's power.

The "wandering" pedestrian explores the urban environment with a body that becomes the interface of the locative technology employed. Its "mission" is envisioned by the method: to "invent" a space without spatial or temporal limits, and is deliberately vague, in the search for new emotions to be found walking around the city streets. Just as in the scenarios of the modern city, sensoriality comes to the forefront of the self, which is a restlessand "arbitrary framework" (Sontag, 1978) of experience.

Significant in this light are Debord's words (2008: 221), "art in the age of its dissolution [...] is at the same time an art of change and the pure expression of the impossibility of change".

The main feature of this type of interactivity – a feature that is worthy of note and which has been only partially analysed – lies in the fact that the actor's intervention is *applicative.* However, this interactivity is open to the great world of entertainment which, from videogames to the activity of *hackers*, is often limited to a puerile hedonistic reward (e.g., breaking into a protected system to create havoc) or else to rudimentary voyeuristic benefits. This is the typical case of pornographic material, which the declared autoerotic purpose confirms as a closed universe, along with the anguished compulsion to remain a prisoner inside it.

None of this is methodologically new for the "frontiers of sociology". Indeed, Simmel's sociology of the senses already spoke of spatial localization as "frame"[69]. The frame sets an artwork apart from the surrounding world, closing it into itself, just as in a society, the relation among the constituent elements acquires spatial expression within the

---

See Debord, G. (1989), *Rapporto sulla costruzione delle situazioni.* Turin, Nautilus; (1998), *La société du spectacle*, Paris, Buchet-Chastel; *In girum imus nocte et consumimur igni*, Milan; (2005), *Panegirico*, Rome, Castelvecchi.

[69] Simmel's approach to the spatial dimension is both *condition* and *symbol* of social relationships. Among the features identified by this German sociologist, "nearness" is the one most closely linked to perception. Nearness is, indeed, a way of coming into relation with others, a way mediated above all by the perception of sensoriality. (cf. Simmel, G. "Lo spazio e gli ordinamenti spaziali della società", in Simmel, G., 1989, *Sociologia*, Mi lan, Edizioni di comunità; cf. also Simmel, G., 1995, *Le metropoli e la vita dello spirito*, ed. Jedlowski P., Milan, Armando; Simmel G., 1984, *Filosofia del danaro* (Cavalli, A., Perucchi, L., ed.), Turin, Utet.

framing *border*. In fact, Simmel sees this border as a way of "carving out" the space which gives social activities their sense.

Simmel and Goffman had already envisioned that spatial experience can be related sensation, to feeling what is happening around one *here and now*. This is the example given by Farman in the idea of the sensory inscribed body. The results are *mobile maps in movement*, samples of which have already been examined. As has already been observed, the temporal dimension that typifies these maps is the *present moment*[70]. This conception is based largely on a phenomenological perspective. In the present moment, "I feel that you feel that I feel". It is from this perspective that affective expression narrates our thoughts and experiences. The same is true for the gestures and movements of others: we can feel ourselves move the same way. We feel this in our bodies and we perceive it in our minds, together. We can even perceive the experience of a whole group of people. Our nervous system is made to "hook up" to that of other humans, so that we can have the experience of others *as ifwe were in their skins*. The history of this "as if", is arranged along the *continuum* of conflicts which, ranging from hair-dying to nose-jobs to sex-changes, has endorsed the passage from loving care of the self to bloody care of the self, almost as if it were written in Braille.

Farman's merit has been to focus on the *spatial orientation* of the body within the environment, in the sensorial body/mobile technology interface. The ability to orient ourselves in space allows us to fix our position in the environment, to estimate distances between external stimuli and to move around. It creates an awareness of the connection between our sensorial body, the spatial context and the circumstance. Tying our shoes, moving around the house, remembering where we left the mobile phone, taking a bottle out of the fridge, and looking for someone in a crowd, are all activities that significantly involve our spatial orientation, even if we are often not aware of this.

What are the differences between our sense of orientation and a GPS navigator? Contrary to the system of this satellite device, our internal system is able to make use of uncertain knowledge, and to test it. Our search for the right street proceeds by trial and error, whereas the GPS is based on sure references, without which it is literally lost. "Turn right in two hundred metres", suggests that suave voice of the computer in the car. We relax and lean back comfortably in our seat and follow the advice, confident of arriving at our destination. But at times, even the most

---

[70] cf, Stern, D.N. (2004), *The Present Moment in Psychotherapy and Everyday Life*.

reliable navigation system fails, maybe because the signal is not clear or the map is not updated. So then we have to go back to the old, difficult job of orienting ourselves in space. But our biological guidance system has a decisive feature: it can use various strategies. In can, for example, make use of approximate information: "I have a feeling I know that street, let's try it". What is more, differently from the GPS, it does not memorize abstract references; instead, it knows what a certain place looks like: "the big red house" or "the tree-lined square". Our orientation system is also able to learn. Someone who knows the street that goes from A to B is usually able to figure out the one from B to C. Even if we are not aware of it, we do not use only visual information to orient ourselves in space. Hearing, smell and touch come into play, and in doing so help the brain to create the cognitive map of the place the body is in. People who can see, for example, find it hard to understand how people blind from birth can, with the sole aid of a cane, get to work on public transportation or enter a restaurant, sit down and pick up knife and fork. But they can in fact do these things, and they do them through recourse to the primary source of information – touch. With the cane they manage not only to identify objects they find along the way, but also to put them in the right position, creating a cognitive map that allows them to know exactly where they are at every moment. The ability to orient oneself in space is above all a matter of practice.

The map is a *complex function* involving diverse areas of the brain. It allows living creatures to orient themselves in surrounding space and to establish their positions in the environment by estimating the distances between external stimuli, and by moving. The spatial map identified in the hippocampus of mice by O'Keefe e Dovstrovsky[71]in 1971, showed that *space constitutes the problem of constraint* par excellence. O'Keefe discovered that when an animal moves its brain, it divides the surrounding environment into many small areas composed like a mosaic. O'Keefe and others demonstrated that the spatial map, that even a simple space, is not formed instantly; rather, it is formed over a period of time ranging from 10 to 15 minutes, from when the mouse enters the new environment.

This characteristic suggests that the formation of the map is a *process of learning.* In his initial formulation, O'Keefe considered the spatial map as a sort of compass used by the animal to orient itself as it "navigates" the territory. Attention to the environment is enough to allow the formation of a spatial map, and its consolidation for a few hours. Long-term stability is

[71] O'Keefe, J. and Dostrovsky,J. (1971), "The Hippocampus as a Spatial Map. Preliminary Evidence from Unit Activity in the Freely-Moving Rat", *Brain Research, 34*, 171-175.

systematically correlated to how much specific attention the environment needs by the animal. When an animal is forced to pay a lot of attention to a new environment, to learn a new spatial task and explore an unknown environment at the same time, the map remains stable for days. It achieves a fixed reality that corresponds to events of sensoriomotor integration. As Pietro Pietrini explains (2007: 69), these are reworked, by a small region in the brain called V5MT, situated in the so-called visual cortex, which gathers information via various sensory paths and then stores them in the hippocampus. The experiment conducted by Japanese researchers on the F4 premotor area of a young macaque, which had already been trained to reach things placed at a certain distance by using a stick or a rake, demonstrates how the *tactile receptive field* varies with the movements of the body in surrounding space. (fig. 4-7).

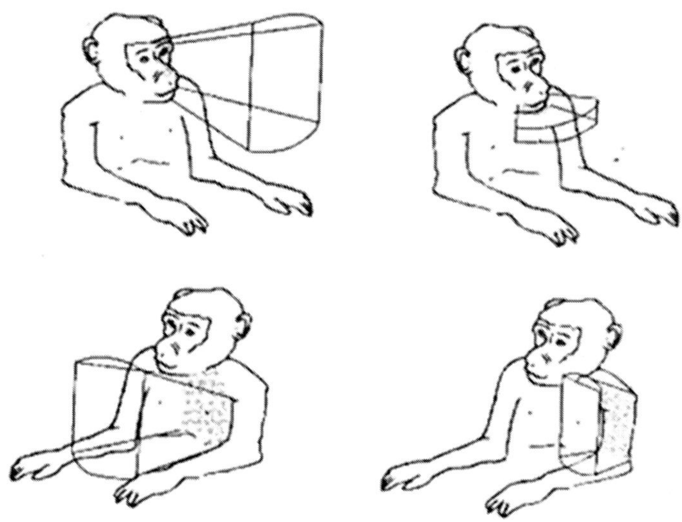

Figure 4-7: Examples of receptive fields of some bimodal neurons present in the F4 premotor area of the macaque. The grey bi-dimensional region indicates the tactile receptive field, while the solid transparent one indicates the corresponding visual receptive field. Fonte: Boncinelli, 2000.

Referring to this experiment, Boncinelli (2000: 173-178) says, as concerns the extension of the tactile receptive field and the corresponding visual field:

> The two fields are spatially connected. The visual field is centred on the tactile, and projects into space starting from the tactile field, for a depth of three or four tens of centimetres. Moreover, the two fields move together if the body part (head or arm) on which the tactile receptive field is situated moves.

As a consequence of this training, the visual field of the bimodal neurons of the animal's F4 premotor area grew by a corresponding length. Some of these neurons know where the animal can arrive now, and its "pericorporeal" space is consequently recalibrated. As the experiment on the young macaque shows, the signalling apparatus found in every part of the body of mammals – in the skin, muscles, retina, and so on – contributes to the construction of neural configurations that make up the map of interaction between the organism and the *Umwelt.*

The construction of these neural configurations is based on the instantaneous selection of neurons and circuits that are involved in the interaction. As Damasio (2007: 387) says in this regard:

> Every brain constructs maps of the environment using its own parameters and internal structure, thereby creating a world specifically linked to the class of brains with a comparable structure.

The spatial map is inseparable from the tactile sensibility of a living body, and in the history of evolution it constitutes the monitoring mechanism that gives each different species "a constantly updated narrative of how we act and why we act" (Humphrey, 1974: 241-255). We could conjecture that the *spatial map* is the *integrated elementary unit* which, from the mammal to *Homo sapiens,* combines the interactive sequences of evolution on the basis of the development of specialized abilities aimed at the solution of specific problems. Indeed, in animal cultures, the spatial map forges the *performances* which, in the biological history of consciousness, found our inheritance of animal intelligence. According to Vallortigara (2000: 410), this inheritance is defined:

> by the borderlines between "me" and "not me", where there are the membranes that delimit organisms and their parts and which govern the exchange of matter, energy and information, events had to be classified aiming at suitable responses. If I like what happens to me, then I stay here; or else I don't like it and then get annoyed and leave.

The majority of interactions between body and environment occur within a *boundary of correlation* that isgenerated by the combination of sensory experience and acting contingency, whichever sense is involved, touch or any other. Today, the light that falls on my body is both something that happens to me ("It burns, I'll move away") and the sign of something "out there" ("The sun is out there, now I know and I'll be able to take advantage of what I know when it's convenient").

The history of this boundary is the same as the history of the relationship between nature and the individual and its biological, technological and sociological transformations. It constitutes the *main property of the living creature.* Without the capacity to continually redefine its boundaries and adapt to them, it would die (Plessner, 2006). Indeed, it is from contexts, understood as experience, that competences which are incomparably more important for a life in continuous evolution, are acquired. For real learning does not consist in storing information and rules. Rather, it consists in the learned freedom to move inside this information and these rules, managing models and at the same time learning how to modify them in order to create the mental ductility that leads to "wisdom" about things and flexibility in their use within the changing boundary that connects technology and the body.

By context we mean a space delimited by the situation within which the dynamics between *challenges* and *skills,* opportunity and ability are played out. Each body finds itself acting in a given situation, using the channel of sensory experience on the basis of circumstance. "Situativity"[72] is the key element that distinguishes modern culture from pre-modern ones.

For a long time, cognitive sciences have given more importance to the vision of action as the execution of predetermined plans, deeming that the social actor is able to control the operations that are necessary on the basis of these plans. In reality, each plan needs to be adapted to the situation in which actors actually find themselves acting.

The main formulation of the theory of *situated action* was made by Lucy Suchman (2006), ethnomethodologist and researcher for the Xerox research centre in Palo Alto (California). The formulation of this theory was chiefly inspired by a series of anthropological studies on the comparison between the navigational techniques of a Pacific people, the Trukesi, with that of Western peoples: whereas the Western navigator starts the voyage with the definition of a route traced on the map, and continues on while trying to keep to that route, the Trukesi – like the Trobrandesi who were analysed by Malinowski – always begin with an objective, and not a plan. The Trukesi start out towards an objective and respond to the situations that present themselves on the way. In proposing

---

[72] The concept of "situativity" of action derives from Soviet psychological culture. See the studies of Vogotskij (1965; 1978), Leont'ev (1959; 1975) and Anokhin (1976).

the organization of the "situated action" as the property coming out of interactions among actors moment by moment (179), Suchman stresses the wide *margin of indetermination* that characterizes the action of a modern Western person, what Symon, Long and Ellis call the "planning paradox"(1996: 12). This margin marks the "court of old truth", which fed its survival on the artificial separation between language and the empirical world, according to the general model of the constitutive rules that construct social reality. This is what Popper, back in the 1930s, called "the positivistic dogma of meaning" [1934: Ital. trans. p.13], which consists in finding the criterion of demarcation between empirical science and metaphysics "in a difference that exists, as it were, in the nature of things" (*ibid*, p.15).

The map aims to capture all the details of an object and an environment, focusing on the contingent construction of use in the interactive exercise of sensory data. The findings of historians and anthropologists show that the use of maps is made reference to in a vast number of different cultural contexts, from the nautical charts of the Marshall islands to the maps depicted by the Australian aborigines (Watson, 1993), and in a space of history that begins with pre-historical times and includes the majority of cultures (Harley, 1987: I).

At the dawn of modernity, the map is "the country that, according to its name, does not exist" (Fiore, 1983: XL43). As Raffaele Itlodeo (the Portuguese sailor who describes Utopia to his listeners in Thomas More's work) explains, in the beginning, the island was a peninsula, that is, literally an "almost-island". At the outset it was something very concrete and not at all ideal. To become what it is at the end required only a slight modification, involving the meeting of what was inside with what was outside. In short, precisely between the two "almost minds" that Peirce speaks about—one whose activity is expressed in concession, that is to say, in the emission of internal data; and the other in interaction with external data. The problem is how to identify the vehicle or instrument that can guarantee their fusion. In the case of sociology, this vehicle is the *body acting in context*: exactly what, since its dawning in the Renaissance, modern culture has considered as the stranger in its philosophical city, at the same time as it is the most honoured guest in its Land of Plenty. It is the secret aim of the sphere of self-referentiality in today's use of the electronic space of "socialisation". A "virtual café" with a 19th- century date.

General Conclusions

What *Homo sapiens* Teaches the Mobile Society of the 21st Century

Our discussion up to now has highlighted the concept of *bodies acting in context* as the main empirical field which allows sociological investigation to test *in loco* the most important neuroscientific discovery of the last two decades, – namely, the existence of mirror neurons, which a group of neurophysiologists led by Giacomo Rizzolatti first identified at the Neuroscience Laboratory of the University of Parma. As Iacoboni says, there is no doubt that "for the first time in history mirror neurons furnish a plausible explanation for complex forms of social interaction"[1]. In Parma, Rizzolatti's team has conducted experimental studies of an area labelled F5, situated in a large cerebral region called the premotory cortex; the portion of the neocortex involved in planning, choosing and performing actions. The F5 area contains millions of neurons which are specialized in the codification of a specific motor behaviour: hand actions like grasping, holding, tearing and, the most fundamental of all, bringing objects

---

[1] Mirror neurons constitute a specific class of neurons that have been discovered in monkeys. Their identification in humans is still underway, but their existence is certain. Neuroimaging studies have revealed that the same neurons activated by a performer during an action, are also activated in an observer of the same action. Further investigations on humans have not only confirmed these neuronal activities on the basis of neuroimaging studies, but they have also led to the conclusion that these neurons are also activated in people with amputated or paralysed limbs when limbs move, as well as in partially or totally blind subjects. For instance, the sound of water being poured out of a pitcher is enough to activate the same neurons in a blind person as those activated in the performer of the action of pouring the water into the glass. Michael Arbib, (2005*) The Mirror System Hypothesis. Linking Language to Theory of Mind,* 2005; Oberman L.M., Pineda J.A., Ramachandran W.S., (2007), "The Human Mirror Neuron System: A Link between Action Observation and Social Skills", *Soc Cogn Affect Neurosci* (2007) 2 (1): 62-66; Rizzolatti G., Sinigaglia C., (2010), "The Functional Role of the Parietofrontal Mirror Circuit: Interpretations and Misinterpretations", *Nature Reviews Neuroscience*, 11 (4) 264-274.

(especially food) to the mouth. For all primates, these actions are basic, as is their production. In the first years of research on mirror neurons, Rizzolatti's team also discovered a group of cells in the F5 area endowed with another characteristic that they could not explain.These are cells that are activated when objects are grasped, and even simply at the sight of graspable objects.. The F5 area contains millions of neurons specialized in the codification of a specific motor behaviour: hand actions like grasping, holding, tearing and, the most fundamental of all, bringing objects (especially food) to the mouth. For all primates, these actions are basic, as is their production. In the first years of research on mirror neurons
The initial studies of the Parma group, which led to the discovery of mirror neurons, came out of Rizzolatti's insights on the role of the premotor areas in creating "spatial maps" around the body. Rizzolatti called this hypothesis "premotor theory of attention".

> The brain that acts is above all a brain that understands!

Rizzolatti and Sinigaglia wrote (2006: 3).

With the discovery of mirror neurons, the human animal discovers itself as *body action in context*. For what is the space of action if not the set of "places" that we can reach by stretching out a hand? Shut up in our sedentary urban existence, we too often forget that even today, for masses of people, the most important activity is what it has been since time began: moving, changing place, walking, etc. On the road, on whatever road, the lesson of *Homo sapiens* comes to life, with tribal chiefs, merchants, ice-cutters, hunters and women bought and sold like market goods, continually losing and finding themselves again. Not to mention animals, whose exploitation as "resources" – in varying ways and degrees among the communities under study – has been the "object" of anthropological study.[2] The classic conflict between those who stress biological determinism in human behaviour and those who, on the contrary, assert the supremacy of ideas and social aspects, has never taken into consideration the possibility that it is, above all, our neurobiology that dictates our social behaviour. As an analysis of sensorimotor changes reveals, in the acts of "seizing with the hand and the mouth" or only "seizing with the hand" the "activities of orientation and of getting hold of" are materialized. As Jean-Pierre Changeux and Paul Ricoeur (1998:162) have underscored, these

[2] Anthropology has always included within the horizon of its studies the interaction between "humans and non-humans", from Frazer's *The Golden Bough* and Durkheim's studies on religion up to the most recent specialised anthropological contributions to Human-Animal Studies.

activities

> contribute to representing the world as a possible environment, full of roads and obstacles, in short to making a liveable world.

Is it possible that mirror neurons can also contribute to anchoring the comprehension of linguistic material to the body and its actions? The philosopher Vittorio Gallese and the cognitive scientist George Lakov were the first to offer this hypothesis in the article, *The Brain's Concepts* (2005).

The existence of mirror neurons reveals the need for a profound change in current ideas about the way our mind works. Undoubtedly, this discovery implies a drastic reappraisal of the model of mind proposed by cognitive psychology, based on a functional analogy with calculators. This approach concentrates above all on defining the formal rules that are supposed to underlie the mind's functioning, completely ignoring the role of bodily experience connected to motor behaviour. On the contrary, mirror neurons imply the existence of a mechanism that allows us to grasp immediately the meaning of other people's actions and even of the intentions behind them, without having to use any kind of reasoning. They effectively provide the physiological explanation for our ability to place ourselves in relation with others. When we observe another being perform a certain action, in our brain, the same neurons that come into play when we personally perform that action are activated. For this reason we can easily understand the actions of others: in our brain, nerve circuits recalling analogous actions that we have performed in the past are activated. This is an extremely important point, for it seems that the "mirror system" comes into play only when social actors are synchronized with a behaviour they have acted out *previously*. For example, it has been observed that for a classical ballet dancer, their mirror neurons are activated exclusively during a performance of classical ballet, and not of modern dance, and vice-versa. The capacity to *learn from experience* is surely the most outstanding aspect of human behaviour. In many ways, we are the personification of what we have learned in the past. In humans, as in other animals, learning and memory have a determining role in almost all our behaviour.

The combinations of the "implicit body" in the execution of a present action represent *Homo sapiens*' salient characteristic, i.e., the tendency to preserve identity through an adaptation to the environment in order to create the conditions for the best possible existence. The physiological, sensory and cognitive mechanisms activated in the relational contexts mediated by current digital technologies are no different from those found in "traditional" relations between humans and technology. The Swedish neuroscientist, Torkel Klingberg (2008: 10-11), has shown that contemporary brain capacity is identical to that of the anatomically modern *Homo sapiens*. What changes, is the creation of the social world. The body of *Homo sapiens*, in part oriented by evolution, in part shaped by environmental interaction, progressively models through trial and error the mobile fabric of the taskscape, or «the sphere of practices and abilities both human and non-human» (Ingold, 1993: 172-174) that punctuate the history of humanity's adaptation to the various local contexts it has met. In the regulation of bodies attested to by tools and artefacts, the appearance of *Homo sapiens* proves that the adaptive individual-environment bond is the common denominator of the synergy which is constantly created between the human brain and the entire organism. This synergy constitutes the organism as an indivisible whole, integrated by the presence of biochemical interactions and neuroregulatory circuits, and involved exclusively as an indivisible whole in the adaptive relationship with the environment. As observation of the cultural architecture of «primarily oral societies»[3]reveals, the work of experience is like a loom, incessantly

[3] The expression «primarily oral culture » was coined by Walter Ong (1982) in order to determine the difference between the cognitive modes of cultures without writing and those of cultures profoundly influenced by its use. According to Ong, the spoken word, as a sound event, is the fruit of a concrete situation and the immediate interaction among human beings. It is a «word-action » that «changes the world ». Word-action against word-memory, change against stasis, memory against forgetting. Analytic thought cannot be communicated, nor even conceived, in a culture without alphabetic writing. Thus, primarily oral cultures have no philosophy. A primarily oral culture transmits knowledge by the spoken word, which is sound. Literate cultures do so mainly through the written or printed word, which is enclosed in a space and perceived by the sight. These two types of culture also have a different relationship with time, as well as with history. The former has no documents, but instead has memory to recall and make things be recalled – an organization of discourse (themes, formulas, proverbs, rhythm), which constitutes a discursive praxis that is also a cognitive feature. With the progressive internalization of writing, which in the Romantic period involved both hand-written and printed culture, there is a change in mnestic economy. Indeed, writing creates distancing between the self and the cognitive object, setting the word in a neutral space – the written page. Once internalized, the sense of analytical precision that is developed along with writing and printing turns back and is reflected onto oral discourse. Consequently, the oral verbalization of a literate person is also different from that of a non-literate person, because their mental structures are different. Ong's book, *Orality and Literacy.The Technologising of the Word* (London and New York, Methuen, 1982, Ital.ed.*Oralità e scrittura*, Bologna, Il Mulino, 1986) represents an important summary of the history of the

weaving past and present together. This poly-functional faculty links human senses with the world in a performance made by a technical instrument. Its progress is rhythmic, like every biological cycle, as recent studies in chronobiology have shown.[4]Its expression is present, like every situation that we experience. Its elaboration is functional to the organization of context, or in other words to the processes that allow people to acquire, codify, store, recall and manipulate information about the nature of their spatial environment.Like every other living creature, human beings are born as bodies that move in interaction with space and the objects that are present in it. They use their senses within a responsive environment by moving in perceptive-motor modes. Their neurobiological architecture develops slowly over time, in accordance with the complexity of thevarying levels of experience that stratify their adaptive networks, in synergy with the context. Experiences can have direct effects on the

---

steps taken by western civilization in its passage from orality to the complete internalization of writing.

[4] Chronobiology (the "study of biological times", from the Greek *chronos*: time) is the discipline that studies biorhythm, or the periodical fluctuations of functions and biological parameters of living organisms. These cycles are also known as biological rhythms. The correlated terms *chronemics* end *chronome* have been used in some cases to describe the molecular mechanisms involved in chronobiological phenomena and the quantitative aspects of chronobiology, particularly when the cycles of different organisms are compared. Chronobiological studies find confirmation in various disciplines -- comparative anatomy, physiology, genetics, molecular biology, etc. On this subject, see Ashikari, M., Higuchi, S., Ishikawa, F., Tsunetsugu, Y., *Interdisciplinary Symposium on "Human Beings and Environments": Approaches from Biological Anthropology, Social Anthropology and Developmental Psychology*, Cambridge University, 2002, http://www.societies.cam.ac.uk/cujif/abstract/020825.htm; Ebert, D., Ebmeier, K.P., Rechlin, T. and Kaschka, W.P. (1998), "Biological Rhythms and Behavior", *Advances in Biological Psychiatry;* Hastings, M. (1998), "The Brain, Circadian Rhythms, and Clock Genes". Clinical review, *British Medical Journal*, 317:1704-1707, http://bmj.bmjjour-nals.com/cgi/content/full/317/7174/1704; Horne, J.A. and Östberg, O. (1976), "A Self- Assessment Questionnaire to Determine Morningness-Eveningness in Human Circadian Rhythms", *International Journal of Chronobiology*, 4, 97-110; ; U.S. Congress, Office of Technology Assessment (1991), *Biological Rhythms: Implications for the Worker, OTA-BA- 463* (Washington, DC, U.S. Government Printing Office, September 1991) http://www.princeton.edu/~ota/disk1/1991/9108/9108.PDF *Biorhythm Experiment Management Plan*, NASA, Ames Research Center. Moffett Field, 1983; *Biological Rhythms and Human Adaptation to the Environment*, US Army Medical Research and Materiel Command (AMRMC), US Army Research Institute of Environmental Medicine.

processes that lead to the development of neural circuits, causing the formation of new synaptic connections and modifying pre-existing ones or favouring their elimination (Kandel, 1998).

When children learn to speak, they do not only learn to associate an object to a determined sound; rather, they connect this sound with a whole set of information, such as the sensations that the object causes to sight or touch, and the emotions it arouses. As they grow, children learn other languages, other names by which to indicate the same object, but this is purely phonetic or orthographic learning. Indeed, though they may achieve good mastery of a foreign language, to the point of blending in with native speakers, their body in action will always give away their true origin.

From the most elementary acts to the most sophisticated ones, requiring a kind ofspecial ability, like performing a dance step or playing the piano, our sensory systems allow our body to correlate its acts with the surrounding world, localizing its interactive consistency. Studies of the somatosensory system (Schacter, Chiu and Ochsner, 1993, 16: 159-182; Schacter, 2001; Diamond, Harris and Peterson, 2001: 183: 195) and of the biology of memory storage (Kandel, 2007) have demonstrated that the corporeal sphere is the *transversal place* of interaction between the sensory dimension and context. It is by now considered certain (Diamond, 2003) that there exists no hierarchy of cognitive functions that would limit sensory systems to processing stimuli "in presence", while systems at supposedly "superior" levels would be the only ones able to perform sophisticated operations of elaboration, codification and retention.

A *sensation*[5] is given when an external stimulus interacts with corporeal receptors. All sensory systems respond to external stimulation according to the following four elementary characteristics: *modality, location, intensity and duration* (Kandel, Schwartz, Jessel: 2000: 428). The differing sensations that we feel and the sensory modes conduct various forms of

[5]I refer here to the well-known studies on the neurophysiological mechanisms of sensation that began in 1925, when Edgar Adrian and Yngwe Zotterman first registered action potential in a sensory nerve that activates the muscle receptor. They discovered that the nerve transmits information from the receptor through a modulation of frequency of electric impulses (Zotterman, 1933, 80: 185: 242). The study of the individual sensory system began with somatic sensation, which is the mode that was the object of the first electro- physiological studies on sensation. Stimulations of somatic sensation are complex, and cover very large areas of the skin. Each different type of receptor is selectively activated by the distinct *spatial properties* of the stimulus. The spatial arrangement of receptors activated in a sense organ conveys important information about the stimulus. In the modes of somatic sensation, the arrangement of the receptors conveys information about the localization of the stimulus both in the body and in the outside world.

energy, which the receptors transduce into electric signals called *potential.* These signals allow human beings to feel many types of events: mechanical, thermal, chemical and electromagnetic. The *location* and spatial dimensions of a stimulus are transported topographically through the activation of the *position* of the receptor in the sensory epithelium called the *receptive field.*

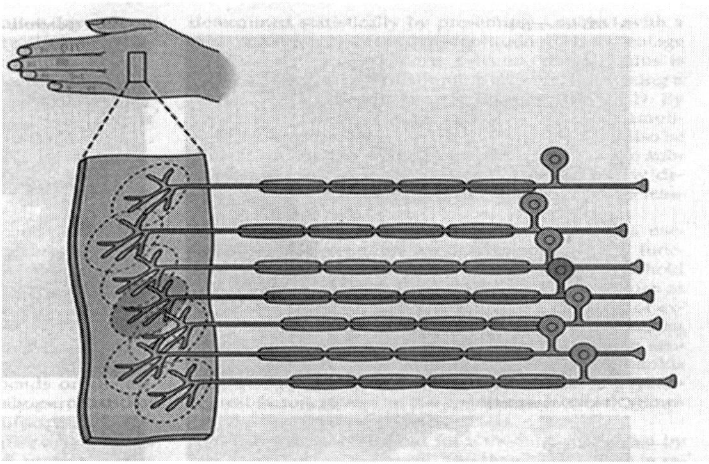

Figure 5-1: Structural basis of the receptive field of receptors for the sense of touch (Kandel, Schwartz and Jessel, 2000: 419)

Bick (1968: 90) experimentally demonstrated the central organising function of skin sensations: actual sensations act as a primary *bodily container*. On the basis of this demonstration, he maintained that the disturbance of primal skin function could lead to the development of a "second skin", through which "object-dependency" would be substituted by a kind of pseudo–independence. What he describes as a "frenetic search for an object" chimes in with the experience of numerous patients, for whom a tattoo or piercingare felt to be imperative, especially in moments of crisis, and as a way of staying together. Based on his work on dermatological conditions, Ulnik (2007) holds that skin mutilation can represent an attempt to maintain the boundaries between body and self. In short, in our culture, the skin is "an envelope of sufferance".

In our view, the origin of this "sufferance" is the "prosthetic rethinking of technology" which, from Kant through to contemporary aesthetics, considers it a distinctive feature of man to delegate the definition of his nature to artificial and supplementary elements, whether a chipped flintstone or a microchip inserted under the epidermis. The philosophical implications of this "prosthetic" conception of technology are, especially following Heidegger, that technology is a particular way of making the world inhabited by humans "appear [to be] a comprehensive environmental project" only partially controllable, since they cannot make it totally transparent. In other words, technology endorses the modern primacy of "representative thought". This means that what man's "distinctive thing" is, from an evolutionary standpoint, *outside of him.* The technological world, originally rising out of an "*externalization*" of the body, takes on growing importance until it represents a universe "endowed with its own life".[6] This poses the embarrassing problem of determining the "game of roles" between the inside (man's faculties) and the outside (technological prosthesis). In other words, is man the "projector" of technology or, on the contrary, "projected" by it? This is the weighty question posed by *E-skin*, a technology that various universities in different parts of the world are involved in creating, with the aim of filling in an important gap in robotics[7].

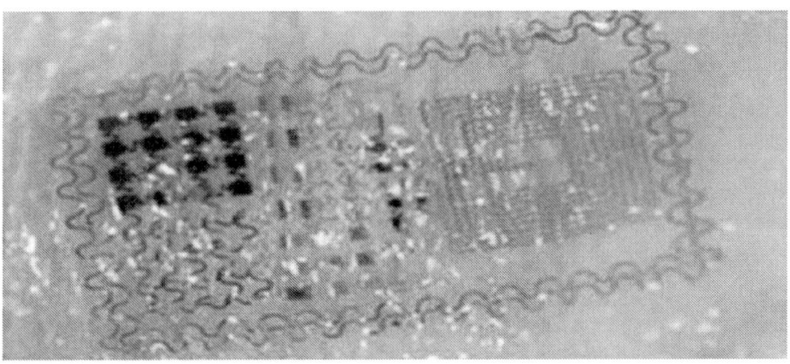

Figure 5-2: http://www.case.edu/artsci/cogs/donald.html.

[6] Cf. Gelhen (1967); Leroi-Gourhan (1977).

[7] John Rogers and his research team at the University of Illinois in Urbana have published EES in the journal *Science*. The EES (Epidermic Electronic System), or more simply *electronic skin*, is thinner than a human hair (it looks like a tattoo) but it has the capacity to restore tactile sensations to people who have lost them, and constantly to register various vital parameters – for example, it can monitor all the data regarding the heart. These researchers stress that the EES could also be used as an electronic bandage to accelerate the healing of wounds and burns. The system comprises a layer of elastic polyester in which miniature sensors, leds, small transmitters and receivers, and a mini-wireless antenna have been integrated. Mini-solar cells provide energy to the system. The device has the same elasticity as human skin and the circuit's creases are perfectly adapted to the skin without obstructing its movements. ("Science", 337, 11 August 2011).

In the worldwide experiment on us, that has been initiated by technology, the most deeply buried root of what it means to be human is being uncovered: the *body acting in context*, which is the foundation upon which the *rational animal* has always rested, and which combines, in the depths of our being, the possibility of reason with the animal nature of the living creature. This body is at stake in portable media devices and mobile communications. The great risk run by knowledge in our times is that the perceptual world might become a puppet theatre, a grand illusion, or even a prison. Life is "now", here and at this moment, and what is done is done. The "new alliance" between human senses and the new media is staged as an "exemplary tale" and "journey" by bodies that move more quickly than originally thought across the maps of contemporary mobile spaces: from the individual's dissolution in social anonymity to the attempt to visibly restore his body in the simulated environments of Interactive Fiction. Nowadays, more and more people spend much of their time playing. They enter large imaginary worlds to defeat dragons and explore unknown realms, they chat with friends using voice chat-lines while trying to hold off alien threats, or they pick up weird plastic guitars and dance wildly around the living-room as if they were on the stage in Madison Square Garden playing to a delirious crowd.

People try to "travel" with their bodies *beyond* the spatial and temporal limits that define the state of reality. The imaginary niche becomes every corner of the world, except for one's own home. In this niche, as in a haunted house or Dracula's castle or a Scottish manor crowded with the ghosts of marketing, people are looking for a "spine-tingling thrill", *as if* they were actually in the presence of the ocean horizon, or a mountain peak above a blanket of clouds, the neon light of a metropolis, the sudden hush of the noise of the forest. In the game of imagination, *homo philosophicus* enjoys the illusion that his moves can forestall the other person's answer in order to reach his own "outside world road"[8]. To quote David Gelernter, computer science expert at Yale, "What is it that people are well-informed about in the age of information? Computer games?"

Games of "pretending" – we need only think of hide-and-seek – can develop capacities learnt better through play than in cases where it is actually necessary to use them. The story of the little boy what cried "Wolf!", heard when we were children, reminds us of what happened to him, without having to learn it in a far harsher way in real life. The more fictional stories we hear, the more circumstances become familiar to us

---

[8] Cf. Bontadini, G. (1996), *Indagini di struttura sul gnoselogismo moderno*, in Studi di filosofia moderna, Milan, Vita e Pensiero.

without actually having to experience them *on our own skin*. "This is the same thing that happened to Sally in that film. What did she do? Oh, yes, that worked pretty well, I think I'll try the same thing".

In the social universe that Freud had to deal with, the most significant obstacle for a balanced psychological life was a crushing Super-Ego. This was the matrix, for example, of sexual inhibitions. A century later, today's pathologies have to do above all with omnipotence, narcissism and the absence of limits. The case of Michael Jackson, the greatest pop star in modern and postmodern history, offers a perfect example here. The most extraordinary, and at the same time disturbing, aspect of his success is the choreography within which Jackson located his voice – a dance that becomes increasingly mechanical, recalling the world of infancy, almost an animated cartoon where the *motor constitution of space* is frozen, as in a cosmetic operation. The flashy choreography of this exceptional Afro-American singer denotes the status of the space around the body in modern culture: "free as a bird" beyond the schemes of the social world, an obsessive territory of pain and loss, albeit rooted in the body and its motor constitution.

The anthropologist of social networks,Danah Boyd, has spoken (2011) about "the collapse of contexts" in reference to the erosion of boundaries and classical divisions in concepts like space, sociality and temporality. The information overload has been with us for some time. Lovink (2011: 38) observes that "the user-patient has to make up for the sensory overload by finding qualitatively important off-line periods".

Tens of studies by psychologists, neurobiologists, educators and web designers arrive at the same conclusion: when we go on-line, we enter an environment that favours distracted, hurried thought and superficial learning. We hit the keys of our PC. We move the mouse, making its wheels go round, and click left or right. We run our fingers over a trackpad. We use our thumbs to write on the small real or simulated keyboards of our blackberryor cell-phone. We scroll our iPhones, iPods and iPads to go from "panorama" to "portrait" visual modes and manipulate icons on their touchscreens.

While we are performing these movements, the "Web" sends a continuous flow of stimuli to our visual, somatosensory and auditory cortices, involving all our senses simultaneously. According to Carr (2011: 146), the Web takes us back to the condition of our origins, when we were distracted by the numerous signals coming "from below", but at the same time it offers many more distractions than our ancestors ever had to deal with. But we need not panic or enter a "pharmacological" mental state, as Bernard Stiegler calls it, in the presence of info-techelegies or mobile story-telling. It would simply be a good idea for us to recognise the progressive, inexorable extinction of the "road of the outside world," which creates the gap of the "missing time" of deepest intimacy with those around us, the time of feelings, confessions and also – to use the term coined by the French economist, Yan Moulier Boutang – of online pollination, with billions of users, like so many bees flying from one site to another, only to increase their owners' value. The screen is black, and there is no other page.

It thus becomes urgent to reconsider seriously "media literacy", not only in terms of the ethics of information, the mastery of gadgets or media know-how, but rather as regards the choice of one's destiny. Perhaps, as Harold Innis, an older colleague of McLuhan's, thought, the real problem of western culture (founded on the Greek-Roman alphabet and the written word) was the loss of the factor of balance found in oral culture: it is the price we paid long ago, when we exchanged the ear for the eye in moving from an oral to a written culture. Merlin Donald, cognitive neuroscientist at Case Western Reserve University, holds that, though humanity is very interested in the changes in the physical ecology of the outer world, we should pay more attention to what is happening inside our bodies (and inside our hearts), when we interact with what surrounds us. Donald foresees that this sudden proliferation of information will probably indicate the direction that our species is going in. Perhaps the next step in the evolution of our information-storing capacity might be to store it internally, once again with the aid of silicone?

Or maybe not. The idea of manipulating our inner nature is disturbing for many of us. What would we do, exactly, with an "expanded" intelligence disconnected with the spatial-temporal dynamics of the human body, like the "rational totality, dominated by rational science"[9]which, according to Husserl, was the "great discovery of modernity"?

There is a great need to revise and enlarge our visions of the human condition, to be closer to reality, not only for others, but above all for ourselves. The *pericorporeal space* revealed by current mobile spaces and experimentally demonstrated by the discovery of mirror neurons, invites cultural forces, and sociology in particular, to a reconsideration of the function of the body in the sphere of its own temporal and spatial systems. According to the acute definition of the American anthropologist, Edward T. Hall (1988: 234), the space surrounding the *body acting in context* constitutes the "hidden dimension" of modern culture. Indeed, a large part

[9] This idea of "mathematised" nature is the cause of the complete change in the idea of world that has come about from Des Cartes on (cf. Husserl, 1961).

of culture lies hidden in the so-called "unconscious", outside the control of the will, though it makes up the woof and warp of human existence. This is the space related exclusively to the interactive events that connect sensory receptors to the *chemical and electromagnetic flows,* structuring here and now the context the body moves in.

From animal to man, they build the bodily architecture through which sensations are transformed into symbols. Numerous studies indicate that our social expectations derive from these more elementary sensory-motor predictions. According to Poincaré (1908: 86-87), these "relations" constitute "potential motor acts" that explain *dynamically* the spatial position of an object. The *motor constitution* of space was defined by Poincaré as the reciprocal coordination of the "multiple parries" made possible simply by the extension of an arm (1913:88). And as far back as 1905 the Austrian physicist Ernst Mach held that, starting from these movements, our body maps the space around it, and that it is by virtue of their purposes that space takes on form for us[10]. Mach affirms that sensations produce external bodies. Thus, every distinction between the physical and psychic element disappears, along with the conventional relation of causality, replaced by the concept of function, defined as the correlation of phenomena.

As Gazzaniga observes (2008: 83), in *pericorporeal space,* the sensomotor complex is the foundation of the constraint of physicality that links us to the environment. The majority of interactions with the environment occur as *localisation,* defined by the body's boundary and corresponding to the irreversible time of the evolutions towards equilibrium that contemporary physics recognises as «rhythmic time of structures whose pulsing is nourished by the flows that cross them" (Prigogine, 1981: 274).

In the harmonic regulation of these flows, *Homo sapiens'* many inventions reveal technology's principle motivation: to create a *network of links* between body and *habitat,* able to integrate is motor constitution with the specificity of the external environment. For instance, studying the

---

[10] The thought of Ernst Mach – particularly his sensationalist interpretation of reality and his criticism of language– directly influenced the Vienna circle and the philosophical current of logical neopositivism, as well as authors like Hermann Bahr, who borrowed from Mach the definition of *Unrettbares Ich* (the unsavable "I"), and Robert Musil, who centred his Ph.D dissertation on Mach's theories. His philosophy is generally known as empirio-criticism (Cf. Mach, E., (1886), *Die Analyse der Empfindungen und das Verhältnis des Physischen zum Psychischen*, Ital. tran. 1975, *L'analisi delle sensazioni* 1903, 1975); (1905), *Erkenntnis und Irrtum*: Ital. tran. 1982, *Conoscenza ed errore*).

work of smiths, Bernstein, a pioneer of movement studies, attributed this process of synchronism to the «secret of dexterity» (Bernstein, 1996).

Evolution has endowed us with a mechanism for a social species – the simple initial nucleus that the relation between sensory information and motor actions arises from. Movement is a fundamental property of living beings which emerged in the course of phylogenic and ontogenetic development, and it is an instrument of learning and social interaction.

Rarely do we stop to appreciate the importance and the refined spatial-temporal architecture of bodily movements. An animal, but also a small child, has to take into account the sensory information that arrives to the brain – an object is near or far, light or heavy, regular or irregular – in order to produce appropriate movements. To slow down a movement of bringing the hand close to an object, grasp it with more or less force according to the estimated weight, regulate the trajectory of the arm movement bringing the object to the mouth or the other hand, are all activities based on the conquest of the final result. We are so used to automatically adjusting our movement in reaching, grasping, moving and so on as to feel that these abilities are limited to relations with the physical world.

The discovery of mirror neurons reveals the intimate, natural and profound bodily link that connects human being with each other and with the elements that here and now comprise their living environment. Consider the smile. A smile passes from one to another, as does laughter. The reaction to a question posed by a polite, smiling person is completely different from the one provoked by the same question asked brusquely. Certain comedians make people laugh just by the quality of their own laughter.

To consider cognitive development as being modulated by a set of biological constraints means recognising the possibility that these constraints can modulate and regulate the learning potential of individuals. At the same time, it can also mean that by determining these potentialities, such constraints have an empowering role, expressed in the incentivization of the individual who learns to channel his/her energies in order to perform a contextualized action (Carey E. Gelman, 1991).

The American psychiatrist Csíkszentmihályi (1997) has used the term flowto define the *focalised concentration* and *absorption* in the work being done. For instance, a painter who is painting a picture and is so absorbed in his/her work as to be unaware of him/herself, or of the passage of time, constitutes a "state of flow". The same state can be reached when a surgeon is performing a difficult operation, in which s/he is obliged to use all his/her skills and capacities. The capacity to respond quickly has therefore increased efficiency and promotes adaptation in a changeable environment. In contrast, uncertain situations have high psycho-physiological costs, as the map below, drawn by Csíkszentmihályi, shows.

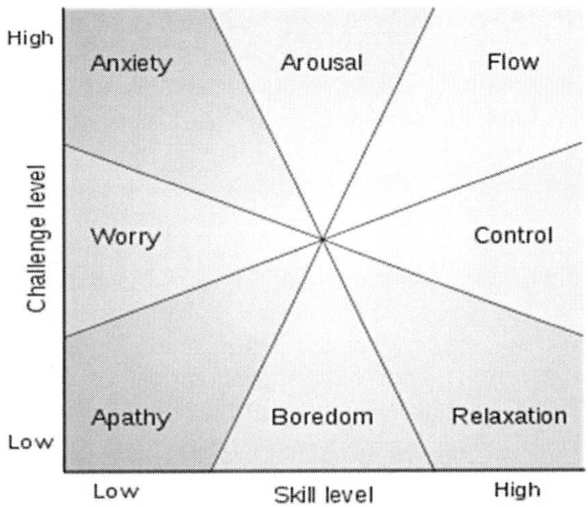

Figure 5-3: Csíkszentmihályi's map (1997), showing how differing mental states can be conceived as products of challenge and skill (1997)

What Csíkszentmihályi has tried to identify are the circumstances that determine the "state of flow". Considering situations in terms of the challenges that each one presents, and the skills of the person involved, the "state of flow" arises in contexts characterised by a high level of challenge and skill, within which the actor's capacity is pitted against the requirements of what s/he wants to achieve. When the challenge is beyond the skill, stress occurs. When skill is superior to the challenge, we acquire a sense of control that can turn into boredom if the level of challenge falls. Emotional responses give birth to this disposition towards action through the involvement of diverse nerve circuits, which in turn lead to a change in the arousalstate of the brain and other systems of the organism. Emotions are processes, usually unconscious, whose essential purpose is to create a state tending towards action and to prepare us to react with determined behaviours to particular environmental stimuli. According to Antonio Damasio[11], the basic emotions are already codified in the trunk of the

[11] Cf. Damasio, A. R. (1994). *Descartes' Error: Emotion, Reason, and the Human Brain.* New York: Avon.; Damasio, A. R., Grabowski, T. J., Bechara, A., Damasio,

encephalon, that is, in a very primitive structure that we even share with reptiles. The primary self does not think, but reacts and feels emotion.[12] The obvious evolutionary advantage associated with the fact that we have base mechanisms that allow us to judge objects and events in the outside world quickly explains the central role of these cerebral processes. The initial warning, "Be careful" in the space of microseconds becomes, "Act!" (Davidson *et al.*, 1990), through arousalprocesses that guide the energy flows inside the system. The *arousal* state activates cognitive mechanisms – "Be careful!" – that do not require conscious participation.

Recent neurological experiments show that the continual use of the arousal stateencourages neurogenesis. The first proof of real growth of the size of brain structures in relation to environmental factors, considered on a large scale, was given only ten years ago, by London taxi-drivers, no less.[13]

The discovery is striking above all for its simplicity and explicative power. Among London taxi drivers, whose job requires them to memorise a large number of complicated routes and localities, the hippocampus is particularly large, larger than for most other people. Considering the importance of the hippocampus in spatial memory, and that London taxi drivers have to memorise such a high number of routes and localities, the conclusion drawn is that their hippocampus is by necessity put under stress, like the muscles of a weight-lifter. Moreover, the hippocampus is larger in the drivers who have been performing this job longer. Its size is, therefore, directly proportional to the number of years spent performing this activity. This suggests a direct relationship between the amount of certain types of mental activity and the size of the neural structure involved in this activity. The results obtained with the taxi drivers are important from several different standpoints. First of all, it was shown that an important neural structure can continue its growth even in adults. Even more relevant is the fact that the growth of a neural structure is *fostered by*

H., Ponto, L. L., Parvizi, J., et al. (2000). "Subcortical and Cortical Brain Activity during the Feeling of Self-generated Emotions".*Nature Neuroscience*, 3, 1049-56.

[12] It has been possible to study experimentally some primary emotions. The findings show that when we observe a manifestation of pain or disgust in others, the same neuron substrate linked to first-person perception of this type of emotion is activated. Another confirmation comes from clinical studies on patients affected by neurological pathologies: once the capacity to feel an emotion is lost, it is no longer possible to recognise that emotion when it is expressed by others.

[13]Maguire E.A. *et al.* (2000), "Navigation-related Structural Change in the Hyppocampi of Taxi Drivers", in *Proceedings of the National Academy of Sciences USA*, 97 (8): 4398- 4403.

its use. The effects of a strong cognitive stimulation in this case seem to win over the deteriorating effects of aging.

Spectacular as they were, for some time the results obtained with taxi drivers were an exception. After all, a single study was not sufficient, precisely because it represented such an extraordinary discovery. A few years later, *Nature*, one of the most highly regarded scientific journals in the world, published a study based on magnetic resonance (MR) and regarding cerebral changes in jugglers.[14]Some volunteers in good health, none of them with previous experience in this activity, were given a three-month training period as jugglers. At the end, the volunteers were able to keep at least three balls in the air. A comparison of their cerebral MRs before and after the training showed an increase in the grey matter in the temporal lobes of both hemispheres, and in the parietal lobe of the left hemisphere. This effects starting declining after the interruption of the practice, and the advantage acquired in the temporal and parietal lobes was reduced. Particularly the third MR exam following interruption of juggling activity revealed this loss.

These concluding observations come together in the "game rule" proven by the "emergence" of *Homo sapiens*: human culture selects elements and rhythms, creates connections and imposes on time its distinctive features like «a musical scale from which particular colours or notes are chosen » (Lewis, 1983: 198). With this "rhythmic rule" *Homo sapiens* bring memorised information online, preparing technology as a *use function* of his genetic patrimony. This "special" capacity has permitted the flexibility and adaptability to differing environments on the basis of the equipment obtained from the territory over time.

And so what can *Homo sapiens* teach the mobile society of the 21st century? Our "time for ourselves", which means feeling at home in one's own body, using past experiences to cope with unexpected risks. Our world is living in an ongoing situation of uncertainty and is, alas, paying too a high price for its increasingly evident incapacity to deal with the events that occur. This time resides in the body and, as John Blaking has to say in the conclusion to *How Musical is Man?*, is waiting to find its full expression and development.

---

[14] Draganski, B. *et al.* (2004), "Neuroplasticity: Changes in Grey Matter Induced by Training", in *Nature*, 427 (6972): 311-312.

BIBLIOGRAPHY

Anzieu D.(1985), *Le Moi-Peau* (1985), Paris, Dunod.

Anokhin, P.K. (1976), *The Philosophical Importance of the Problem of Natural and Artificial Intellects*, in "Soviet Studies in Philosiphy", 14 (4), pp. 3–27.

Arbib, Michael. (2005), *The Mirror System Hypothesis. Linking Language to Theory of Mind,* 2005, retrieved 2006-02-17.

Ashikari, M., Higuchi, S., Ishikawa, F., Tsunetsugu, Y. (2002), *Interdisciplinary Symposium on "Human Beings and Environments": Approaches from Biological Anthropology, Social Anthropology and Developmental Psychology*, Cambridge University, http://www.societies.cam.ac.uk/cujif/abstract/020825.

Assman, A. (2006), "The Printing Press and the Internet: from a Culture of Memory to a Culture of Attention." In *Globalization, Cultural Identities and Media Representations*, edited by Natasha Gentz and Stefan Kramer, 11-23. Albany: State University of New York Press.

Augé, M. (1995), *Non–Places: Introduction to an Anthropology of Supermodernity*, London, Verso.

—. (2003) *Le Temps en Ruines. Paris, Galilée.*

Bachelard, G. (2000), *The Dialectic of Duration*, Manchester, Clinamen Press.

Baddeley, A. Eysenck M.V., Anderson M.C., (2009), *Memory*, New York, Psychology Press.

Barthes, R. (1981), *Camera lucida*, New York, Hill & Wang.

—. (1982), *Empire of Signes*, New York, Hill & Wang.

Bataille, G. (1962), *Death and Sensuality. A Study of Eroticism and the Taboo*. New York, Walker & Co.

Bartlett, F.C. (1932), *Remembering: a Study in Experimental and Social Psychology*, New York, Cambridge University Press.

Baudrillard, J. (1975), *The Mirror of Production*, New York, Telos Press.

—. (1981), *For a Critique of a Political Economy of the Sign*, St Louis, Telos Press.

—. (1983), *Simulations,* New York, *Semiotext (e).*

—. (1990), *Seduction*, London, Mac Millan.

—. (1993), *Symbolic Exchange and Death*, London, Sage Publications &TCS.

Bauman, Z. (1981), "Semiotics and the Function of Culture", in Julia Kristeva et al. (eds.), *Essays in Semiotics*, The Hague / Paris, pp.279–291.

Benjamin, W. (1969), *The Work of Art in the Age of Mechanical Reproduction.*in Hannah Arendt (ed.), *Walter Benjamin: Illumi-nations, Essays and Reflections,* New York, Schocken Books.

Bergson, H. (1896), *Matière et Mémoire. Essai sur la relation du corps à l'ésprit* ; ed.it. (pessina A., a cura di), *Materia e memoria. Saggio sulla relazione tra il corpo e lo spirito*, Roma–Bari, Laterza.

Berry, C., Kim, S., Spigel, L. (eds). (2010), *Electronic Elsewheres. Media, Technology, and the Experience of the Social Space*, Minneapolis, University of Minnesota Press.

Bick E. (1968), "The Experience of the Skin in Early Object Relations", *International Journal of Psychoanalysis*, 1968, 49: 558–566; trad.it.in Bonamibo, V., Iaccarino, B., *L'osservazuione diretta del bambino,* Torino, Bollati Boringhieri.

Boltanski, C. (1995), *La présence des absents*, Les cahiers de la Villa Gillet, n° 2, 51-71.

Bontadini, G. (1996), *Indagini di struttura sul gnoselogismo moderno*, in Studi di filosofia moderna, Milano, Vita e Pensiero.

Boyd, D. (2011), "Social Network Sites as Networked Publics. Affordances, Dynamics, and Implications" in Papacharissi, Z. (ed.), *A Networked Self. Identity, Community, and Culture on Social Network Sites,* New York, Routledge.

Borkenau, F. (1984), *La transizione dall'immagine feudale all'immagine borghese del mondo. La filosofia del periodo della manifattura*, Bologna, Il Mulino.

Bourdieu, P. (1977), *An Outline of a Theory of Practice*. Cambridge, Cambridge University Press.

—. (1984), *Distinction.A Social Critique of the Judgement of Taste.* Cambridge, MA. Harvard University Press.

Blustein, J. (2008), *The Moral Demands of Memory*. Cambridge-New York: Cambridge University Press.

Bolter J.D. (1991), *Topographic Writing: Hyperetxt and the Electronic Writing Space,* in *Hypermedia and Literary Studies*, edited by Delany & Landow, Cambridge (Mass), the MIT Press.

Bruce, D. (1995), "The How and Why of Ecological Memory", *Journal of Experimental Psychology*, 77: 114: 78-90.

Cardona, G.R. (1981), *Antropologia della scrittura*, Torino, Loescher.

Carpenter, C.R. (1958), "Territoriality : "A Review of Concepts and Problems", *i*n Roe, A. and Simpson G.G. (eds.), *Behavior and*

Evolution, New Haven, Yale University Press.

Carr, N. (2010), *The Shallows. What the Internet is doing to our Brains*; trad.it. *Internet ci rende stupidi? Come la Rete sta cambiando il nostro cervello,* Milano, Raffaello Cortina editore.

Cassirer, E. (1923–29), *Philosophie der symbolischen Formen*; trad.inglese (1953–57), *Philosophy of Symbolic Forms.*

Castells, M. (1996), *The Rise of the Network Society, The Information Age: Economy, Society and Culture,* Vol. I. Cambridge, MA; Oxford, UK. Blackwell.

—. (1997), *The Power of Identity, The Information Age: Economy, Society and Culture,* Vol. II. Cambridge, MA; Oxford, UK. Blackwell.

—. (1998), *End of Millennium, The Information Age: Economy, Society and Culture,* Vol. III. Cambridge, MA; Oxford, UK. Blackwell.

—. (2008), *Mobile communication e trasformazione sociale*, Milano, Guerini e Associati.

—. (2009), *Comunicazione e Potere*, Milano, Bocconi Università Edizioni.

Castells, M. &Pekka Himanen, P. (2006), *The Information Society and the Welfare State: The Finnish Model.*Oxford UP, Oxford, 2002; trad.it. *Società dell'informazione e Welfare State. La lezione della competitività finlandese*, Milano, Guerini e Associati.

Changeux, J.P., Ricoeur, P. (1998), *La nature et la regle*, Paris, Odile Jacob ; trad..it. 1999, *La natura e la regola. Alle radici del pensiero*, Milano, Cortina.

Cohen, G. (2008), "The Study of Everyday Memory", in *Memory in the Real World*, New York, Psychology Press, pp. 1-20.

Collins, R. (1988), *Theoretical Sociology*, New York, Harcourt Brace Janovich Publishers.

Cosmides, L. & Tooby, J.(1994), "Beyond Intuition and Instinct Blindness: Toward an Evolutionarily Rigorous Cognitive Science". *Cognition,* 50 (1-3), 41-77.

Crang, M., Graham, S. (2007), "Sentient Cities, Ambient Intelligence and the Politics of Urban Space", in *Information, Communication & Society*, 10 (6), pp. 789– 817.

Cranshaw, J., Toch, E., Hong, J., Kittur, A., Sadeh, N. (2010), "Bridging the Gap between Physical Location and Online Social Networks", in *Proceedings of Ubicomp* '10.

Crow, B., Longford, M., Sauchuk, K. (2010), *The Wireless Spectrum*, Toronto, University of Toronto.

Csíkszentmihályi, M. (1997), *Finding Flow: The Psychology of Engagement with Everyday Life,* New York, Basic Books.

Damasio, A. (1994), *Descartes'Error: Emotion, Reason, and the Human*

Brain, New York, Putnam: trad.it. (1995), L'*errore di Cartesio*, Milano, Adelphi.
—. (1999), *The Feeling of What Happens: Body and Emotion in the Making of Consciousness*, Orlando, Harcourt; trad.it. (2000), *Emozione e coscienza*, Adelphi.
—. (2000), *Looking for Spinoza: Joy, Sorrow, and the Feeling Brain*, Orlando, Harcourt; trad.it. (2003), *Alla ricerca di Spinoza.Emozioni, sentimenti e cervello*, Milano, Adelphi.
—. (2010), *Self Comes to Mind: Constructing the Conscious Brain,* New York, Pantheon; trad.it.*Il sé viene alla mente. La costruzione del cervello cosciente.*
Debord, G.(1989), *La Société du spectacle*, Paris, Èditions Buchet-Chastel,
—. (1998), *In girum imus nocte et consumimur igni*, Milano, Mondadori.
—. (1999) *Rapporto sulla costruzione delle situazioni.* Torino, Nautilus.
—. (2005) *Panegirico*, Roma, Castelvecchio.
De Certau, M. (2001), *L'invention du quotidien. Arts de faire*, Paris, Gallimard.
De Kerckhove, D. (2003), *La conquista del tempo*, Roma, Editori Riuniti.
Deleuze. G. Guattari, F. (1972), *L'Anti-Edipo*, volume I di *Capitalismo e schizofrenia*, tr. Alessandro Fontana, Einaudi, Torino 1975.
—. (1975), *Kafka. Per una letteratura minore* tr. Alessandro Serra, Feltrinelli, Milano 1975 e Quodlibet, Macerata 1996.
—. (1976), *Rizoma* tr. Stefano Di Riccio, Pratiche, Parma 1977.
—. (1980), *Millepiani* volume II di *Capitalismo e schizofrenia*, tr. Giorgio Passerone, Istituto della Enciclopedia Italiana, Roma 1987.
Descartes, R.(1637–1968), *Discourse on Method and Meditations*, Harmondsworth, Penguin.
Detienne, M. (1963), *La notion de daimon dans le pythagorisme ancien*
—. (1963), *De la pensée religieuse à la pensée philosophique*
—. (1967), *Les Maîtres de vérité dans la Grèce archaïque*, trad. it. Augusto Fraschetti, *I maestri di verità nella Grecia arcaica*, Bari-Roma: Laterza 1977.
—. (1974), *Les ruses de l'intelligence: la métis des Grecs*, (in collaborazione con Jean-Pierre Vernant, trad. it. Andrea Giardina, *Le astuzie dell'intelligenza dell'antica Grecia*, Bari-Roma: Laterza 1984.
—. (1977), *Dionysos mis à mort* (trad. it. Mario De Nonno, *Dioniso e la pantera profumata*, Bari-Roma: Laterza 1983).
—. (1977), *La cuisine du sacrifice en pays grec*, (in collaborazione con Jean-Pierre Vernant e con scritti di Jean-Louis Durand e altri, trad. it. *La cucina del sacrificio in terra greca*, Torino: Bollati Boringhieri, 1982).

—. (1980), *Mito/rito* in *Enciclopedia Einaudi*, Torino: Einaudi 1980, vol. 9, pp. 348-363.
—. (1981), *L'invention de la mythologie* (trad. it. Flavio Cuniberto, *L'invenzione della mitologia*, Torino: Bollati Boringhieri 1983).
—. (1986), *Dionysos à ciel ouvert* (trad. it. Maria Garin, *Dioniso a cielo aperto*, Bari-Roma: Laterza 1987 e 2000).
—. (1988), *Les Savoirs de l'écriture en Grèce ancienne* (in collaborazione con Giuseppe Cambiano e altri, trad. it. Maria Rosaria Falivene, *Sapere e scrittura in Grecia*, Bari-Laterza: Laterza 1997).
—. (1989), *L'écriture d'Orphée*, (trad. it. Maria Paola Guidobaldi, *La scrittura di Orfeo*, Bari-Roma: Laterza 1990).
—. (1989), *La vie quotidienne des dieux grecs* (in collaborazione con Giulia Sissa, trad. it. Claudia Gaspari, *La vita quotidiana degli dei greci*, Bari-Roma: Laterza 1989 e Milano: Mondadori 1994)
—. (2005), *Les Grecs et nous* (trad. it. Arianna Ghilardotti, *Noi e i greci*, Milano: Cortina, 2007).
De Souza & Silva, A., Sutko D.M. (2009), *Digital Cityscpaes. Merging Digital and Urban Playspaces*, New York, Peter Lang.
De Souza & Silva, A., Frith, J. (2010), " Local Mobile Social Networks. Mapping Commu-nication and Location in Urban Spaces", *Mobilities,* 5 (4), pp, 485– 506.
De Kerckhove, D. (2003), *La conquista del tempo*, Roma, Editori Riuniti.
Descola, Ph. (2005), *Par–de–là nature et culture*, Paris, Gallimard.
Diamond, M.E. (2002a), "Percezione, apprendimento, corteccia cerebrale: il passato, le nuove scoperte e le prospettive future", in Fraun-felder, E., Santoianni, F. (a cura di), *Le scienze bioeducative, Prospettive di ricerca*, Napoli, Liguori.
Draganski, B., Gaser C., Busch *et al.* (2004), "Neuroplasticity: Changes in Grey Matter Induced by Training", *Nature* 427.
Dugger, R.L.C. (2002), *Itea/Gallup Poll Reveals What Americans Think about Technology.*
Dunbar, R.I.M. (2004), *The Human Story*, London, Faber abd Faber.
Durkheim, E. (1971), *Le forme elementari della vita religiosa*, Milano, Edizioni di Comunità.
—. (1973), *Rappresentazioni individuali e rappresentazioni collettive*, Roma, Newton Compton Italiana.
Eagle, N. (2010), « Mobile Phones as Social Sensors », in *The Handbookof Emergent Technologies in Social Research*, Oxford, Oxford University Press.
Edelman, G.M. (1987), *Neural Darwinism*, New York, Basic Books; trad.it.*Darwinismo neurale.La teoria della selezione dei gruppi*

neuronali. Torino, Einaudi.

—. (1989), *The Remembered Present*, New York, Basic Books; trad.it.1991, *Il presente ricordato*, Milano, Rizzoli.

—. (1992), *Bright Air, Brilliant Fire.*

Edelman G.M., Tononi, G.(2000), *A Universe of Consciousness.How Matter becomes Imagination*; trad.it.*Un universo di coscienza. Come la materia diviene immaginazione*, Torino, Einaudi.

Ebert, D., Ebmeier, K.P., Rechlin, T. e Kaschka, W.P. (1998), "Biological Rhythms and Behavior", *Advances in Biological Psychiatry.*

Ekstrom, S. (2004), 'The Mind beyond our Immediate Awareness: Freudian, Jungian, and Cognitive Models of the Consciousness' in: *Journal of Analytic Psychology*, 2004, 49: 657-682.

Durlach N.I., Mavor A.S. (1995), "Committee on Virtual Reality. Research and Development", *Virtual Reality.Scientific and Technological Challenges*, National Research Council of United States.

Farinelli, F. (2003), *La crisi della ragione cartografica*, Torino, Einaudi.

Farman, J. (2011), *Mobile Interface Theory: Embodied Spaces and Locative Media,* New York, Routledge.

Foucault, M. (1975), *Surveiller et punir : Naissance de la prison*, Paris, Gallimard.

Francastel, P. (1974), Peinture et société : naissance et destruction d'un espace plastique de la Renaissance au cubisme, Paris, Lyon, Audin.

—. (1968), *Art et technique aux XIXe et XXe siècles*, Paris, Gallimard, 1968 (Tel) ;

—. (1978), *La Réalité figurative : éléments structurels de sociologie de l'art*, Paris, éd. Gonthier.

Freud, S. (1895), *Progetto per una psicologia scientifica* in *Opere*, (1976-1980), Torino, Bollati Boringhieri, vol. II.

—. (1898), *L'interpretazione dei sogni*, in *Opere*, (1976-1980), Torino, Bollati Boringhieri, vol. III.

—. (1901), *Psicopatologia della vita quotidiana*, in *Opere*, (1976-1980), Torino, Bollati Boringhieri, vol. IV.

—. (1907), *Delirio e sogni nella "Gradiva" di Jensen*, in *Opere*, (1976-1980), Torino, Bollati Boringhieri, vol. V.

—. (1913), *Totem e tabù*, in *Opere*, (1976-1980), Torino, Bollati Boringhieri, vol. VII.

—. (1914), *Introduzione al narcisismo*, in *Opere*, (1976-1980), Torino, Bollati Boringhieri, vol. VIII.

—. (1917), *Lutto e melanconia*, in *Opere*, (1976-1980), Torino, Bollati Boringhieri, vol. VIII.

—. (1919), *Il perturbante,* in *Opere*, (1976-1980), Torino, Bollati

Boringhieri, vol. IX.
—. (1920), *Al di là del principio del piacere* , in *Opere*, (1976-1980), Torino, Bollati Boringhieri, vol. IX.
—. (1927), *L'avvenire di un'illusione*, in *Opere*, (1976-1980), Torino, Bollati Boringhieri, vol. X.
—. (1929), *Il disagio della civiltà*, in *Opere*, (1976-1980), Torino, Bollati Boringhieri, vol. X.
Fuster, .M.(1997) *The Prefrontal Cortex: Anatomy, Physiology and Neuro-psychology of the Frontal Lobe*, 3rd ed. Philadelphia: Lippincott-Raven.
Gallese V., Lakoff.(2005), "The Brain'sConcepts: The Role of the Sensory-motor System in Reason and Language Cognition", in *Neuro-psychol.*, 22 , pp. 455–479.
Gardner, H. (1987), *Formae mentis. Saggio sulla pluralità dell'intelligenza*, Milano, Feltrinelli.
Gazzaniga, M. (2008), *Human,* ; trad.it. *Human. Quel che ci rende unici* Milano, Raffaello Cortina editore.
Geertz, C. (1973), *Interpretation of Cultures*, New York, Basic Books.
Gehlen, A. (1957); trad.it. 1967, *L'uomo nell'era della tecnica*, Milano, Sugar editore.
Gibson, J.J. (1979), *The Ecological Approach to Visual Perception*, Hillsdale, NJ, Herlbaum.
Giddens, A. (1979b), *Cenral Problems in Social Theory: Action, Structure & Contradiction in Social Analysis*, London, MacMillan.
—. (1990b), *The Consequences of Modernity*, London, Polity Press.
Goffman, E. (1986), *Frame Analysis. An Essay on the Organization of Experience,* Boston, Northeastern University Press.
Goldberg, E. (2009), *The Executive Brain: Frontal Lobes in a Complex World*, New York, Oxford University.
Goldman-Rakic P.S. (1987), "Circuitry of primate prefrontal cortex and regulation of behavior by representational memory", in *Handbook ofPhysiology: Nervous System Higher Functions of the Brain*, Part 1, Bethesda MD: American Physiological Association, 373: 417.
Goodall, J. (1968), *The Chimpanzees of Gombe: Patterns of Behavior*, Cambridge, Mass., Harvard University Press.
Gould, J.L. (1984), "Natural History of Honeybee Learning" in *The Biology of Learning* (eds. Marler, P. &Terrace H.S.), Berlin, Springer Verlag, pp. 149–180.
—. (1986), *The Locale Map of Honey–bee: Do insects have Cognitive Maps*, in "Science", 232, pp.861–863.
Gould J.L. & Marler, P. (1987), *Apprendimento e istinto*, in "Le Scienze",

223, pp. 29–38.

Gould S.J. (1981), *The Mismeasure of Man*, New York, W.W. Norton & Co; trad it.*Intelligenza e pregiudizio. Le pretese scientifiche del razzismo.*

Gould, S.J. & Vrba, E.S. (1982), *Exaptation. A Mixing Term in the ScienceofForm*, in “Paleobiology”, 8, pp.4–15.

Gruzinski, S. (1994), *La colonnizzazione dell'immaginario. Società indigene e occidentalizzazione nel Messico spagnolo*, Torino, Einaudi.

Halbwachs, M. (1992), *On Collective Memory,* edited, translated and with an introduction by L.A. Coser. Chicago–London, The Chicago University Press.

Hall, E.T. (1959), *The Silent Language*, Garden Garden City, New York; trad.it.1966 *La dimensione nascosta.Vicino e lontano. Il significato delle distanze tra le persone*, Milano, Bompiani.

—. (1984), *The Dance of Life. The Other Dimension of Time*, New York, Anchor Books.

Hamilton, W.D. (1964), *The Genetical Theory of Social Behavior*, in “Jour-nal of Theoretical Biology”, 7, pp.1–52.

Henshilwood, C.S., D'Errico, F., Yates, R., Jacobs, C., Tribolo, C., Duller, G.A.T., Mercier, N., Sealy, J.C., Valladas, H., Watts, I., Wintle, A.G. (2002), *Emergence of Modern Human Behavior: Middle Stone Age Engraving from South Africa,* in “Science”, 295: pp. 1278–1280.

Hastings, M. (1998), “The Brain, Circadian Rhythms, and Clock Genes”. Clinical Review, *British Medical Journal*, 317:1704-1707, http://bmj.bmjjournals.com/cgi/con tent/full/317/7174/1704

Horne, J.A. e Östberg, O. (1976), “A Self-Assessment Questionnaire to Determine Morningness-Eveningness in Human Circadian Rhythms”, *International Journal of Chronobiology*, 4, 97-110.

Humphrey, N.K. (1983), *Consciousness Regained*, Oxford, Oxford University Press.

—. (1992), A *History of the Mind*, London, Chatto&Windus; trad it.*Una storia della mente*, Torino, Instasr, 1998.

Hebb, D.O. (1949*), Organization of Behavior*, New York, Wiley; trad.it. 1975, *L'organizzazione del comportamento*, Milano, Angeli.

Husserl E. (1961), *La crisi delle scienze europee e la fenomenologia tra-scendentale*, Milano, Il Saggiatore.

—. (1931), *Méditations cartesiennes*, Paris, PUF.

Iacoboni, M. (2008), *Mirroring People. The New Science of How we connect with Others*, trad.it. *I neuroni specchio. Come capiamo ciò che fanno gli altri,* Torino, Bollati Boringhieri.

Ingold, T. (1990), An Anthropologist Looks at Biology, in *Man (N.S.)*, vol. 25, n. 2, 208-229.

—. (1993b), *The Temporality of the Landscape*, 'World Archaeology' 25:152-174.

—. (1993a), "Tool Use, Sociality and Intelligence", in Gibson K. and Ingold T. (eds.) *Tools, Language and Cognition in Human Evolution*, Cambridge: Cambridge University Press, 429-445.

—. (1993b), "Tools and Hunter-Gatherers", in Berthelet A. and Chavaillon J. (eds.) *The Use of Tools by Human and Non-Human Primates*, 281-295, Oxford: Clarendon Press.

—. (1997), Eight themes in the anthropology of technology, in *Social Analysis*, n. 41(1), 106-138.

—. (2000), *The Perception of the Environment: Essays on Livelihood, Dwelling and Skills*, London-NY: Routledge.

—. (2006), "Walking the Plank: Meditations on a Process of Skill", in J. R. Dakers (ed), *Defining Technological Literacy: Towards an Epistemological Framework*, New York: Palgrave Macmillan, pp 65-80.

James, W. (1907/1995), *Pragmatism*. New York: Dover Publications.

—. (1890/1981), *The Principles of Psychology*. Cambridge, Mass., Harvard University Press.

Jenkins, H. (2006), *Convergence Culture*, New York, New York University.

Kagan, J. & Baird, A. (2004), "Brain and Behavioral Development during Child-hood", in M. Gazzaniga (ed.), *The Cognitive Nneuroscience III*, Cambridge, Mass., MIT Press, pp. 93: 103.

Kay, A. (1996), "Revealing the Elefant: The Use and Misuse of Computer in Education", *Sequence.* Vol. 3, No. 4 (Jul./Aug.) p. 22-28.

Kandel, E. (2007), *In Search of Memory*: *The Emergence of a New Science of Mind*, New York, W.W. Norton&Company.

Kelly, K. (2010), *hat Technology Wants*, New York, Viking Penguin.

Klawans, H.L. (1988), *Toscanini's Fumble and Other Tales of ClinicalNeurology*. Chicago-New York: Contemporary Books.

Klingberg T. (2008), *The Overflowing Brain*: *Information Overload andthe Limits of Working Memory*, New York: Oxford University Press.

Lacan, J. (1953), *Towards a Genetic Theory of the Ego*, in "Some Reflections on the Ego", *The International Journal of Psycho-Analysis*, vol. 34: 12-15.

Lakoff, G. & Johnson M. (1999), *Philosophy in the Flesh: the Embodied Mind and its Challenge to Western Thought*, New York, Basic Books.

Latour, B. (1991), *Laboratory Life: the Social Construction of Scientific*

Facts, Sage, Los Angeles, USA.

—. (1987), *Science In Action: How to Follow Scientists and Engineers Through Society*, Harvard University Press, Cambridge Mass., USA.

—. (2005), *Reassembling the social: an introduction to Actor-network theory*, Oxford; New York, Oxford: University Press.

Lefebvre, H. (1991), *The Production of Space*, Oxford, Blackwell.

—. (2004), *Rhythmanalysis, Time and Everyday Life*, London, Continuum.

Lemma, A.(2005), *Under the Skin, A Psychoanalytic Study of Body Modification*, London, Routledge. Ed. it.(2011) *Sotto la pelle*, Milano, Raffaello Cortina.

Leroi–Gourhan, A. (1964) (2 voll.); trad.it. (1977), *Il gesto e la parola*, Torino, Einaudi.

Lévi-Strauss, C. (1962), *La pensée sauvage*, Plon, Paris, Plon; trad.it. (2003), *Il pensiero selvaggio,* Milano, Il Saggiatore.

Lewis, G.(1980), *Day of Shining Red* .Cambridge University Press; trad.it. (1984), *Giorno di rosso splendente: saggio sul rituale,* Milano. Franco Angeli.

Lieberman P. (1994), "The Origins and Evolution of Language", in Ingold T. (ed.), *Companion encyclopedia of anthropology*, London: Routledge: 108-132.

LeDoux, J. (2002), *Synaptic Self: How Our Brains Become Who We Are*, New York, Viking Penguin.

Le Goff, J. (1992), *History and Memory*.translated by S. Randall and E.Claman. New York: Columbia University Press.

Lewin, R. (1999), *Human Evolution*, London, Blackwell Science.

Leont'ev, A.N. (1977), *Attività, coscienza, personalità*, Firenze, Giunti.

LLinás, R.R. (2002), *I of the Vortex. From Neurons to Self* , Cambridge, Mass: the MIT Press.

Lorenz, K. (1935), *Der Kumpan in der Umwelt des Vogels*, in " Journal of Ornithology", 83, pp. 137–213; trad.ingl. *Studies in Animal and Human Behaviour,* Cambridge (Mass.), harvard University Press, 1970.

—. (1949), *Er redete mit dem Vieh, den Vögelln und den Fischen*, Wien, Borotha– Schoeler; trad.it. *L'anello di Re Salomone*, Milano, Adelphi, 1994.

—. (1965), *Evolution and Modification of Behavior*; trad.it. *Evoluzione e modificazione del comportamento*, Torino, Bollati Boringhieri.

—. (1973), Die Rückseite der Spiegels, München, Piper; trad.it. *L'altra faccia dello specchio*, Milano, Adelphi, 1982).

Lovink, G. (2012), *Netwoks without a Cause*, New York, Polity Press; trad.it.

—. *Ossessioni collettive. Critica dei social media*, Milano, Egea.
Luckmann Th. (1993), *Remarks on Personal Identity: Inner, Social, and Historical Time,* in A. Jacobson-Wigging (ed.by) ,*Identity: Personal and Socio-cultural. A Symposium*, Almqwist & Wiksell, Uppsala.
Lurija, A.R., L.S.Tsvetkova L.S., Futer, D.S. (1965), "Aphasia in a Composer", in *Journal of Neurological Science*, 2, pp. 288–292.
Mach, E.(1996), *Die Analyse der Empfindungen und das Verhältnis des Physischen zum Psychischen*, Jena, Verlag von Gustav Fischer ; trad.it. *L'analisi delle sensazioni e il rapporto tra fisico e psichico*.
Maffesoli, M.(1979), *La Conquête du présent. Pour une sociologie de la vie quotidienne*, Paris, PUF.
—. (1991), *L'Ombre de Dionysos*, Le Livre de Poche.
—. (1991), *Le Temps des tribus*, Paris, Le Livre de Poche.
—. (1996), *La Contemplation du monde*, Le Livre de Poche.
—. (1996), *Éloge de la raison sensible*, Paris, Grasset.
—. (1997), *Du nomadisme. Vagabondages initiatiques*, Paris, Le Livre de Poche.
—. (2003), *L'Instant éternel. Le retour du tragique dans les sociétés postmodernes*, Paris, La Table Ronde.
—. (2007), *Le Réenchantement du monde. Morales, éthiques, déontologies*, Paris, éd. Table Ronde, 2007.
—. (2008), *Iconologies. Nos idolatries postmodernes*, Paris, Albin Michel, 2008.
Maldonado, T.(2005), *Reale e virtuale*, Milano, Feltrinelli.
Marler, P. (1970), "A Comparative Approach to Vocal Learning: Song Development in White–Crowned Sparrows" in *Journal of Comparative and Physiological Psychology*, 71, pp. 1–25.
Mauss, M. (1990), *The Gift: Forms and Functions of Exchange in Archaic Societies*. London, Routledge; trad.it. (2002), *Saggio sul dono. Forma e motivo dello scambio nelle società arcaiche*, Torino, Einaudi.
Mazzoli, L.(2009), *Network Effect. Quando la rete diventa pop.*
Mc Gonigal, J. (2011), *Reality Is Broken: Why Games Make Us Better and How They Can Change the World.*
Mc Luhan, M. (1962), *The Gutenberg Galaxy: the Making of Typographic Man*, Toronto, University of Roronto Press; trad.it. (1976), *La Galassia Gutenberg. Nascita dell'uomo tipografico*, Roma, Armando.
Mead, G.H. (1934), *Mind, Self and Society*, trad. it. 1966 , *Mente, sé e società*, Firenze, Giunti Barbera.
Merleau–Ponty, M. (1942), *La Structure du comportement*, Paris, PUF
—. (1945), *Phénoménologie de la Perception*, Paris, Gallimard.
Mithen, S. (2007), *The Singing Neanderthals. The Origins of Music, Lan-*

guage, Mind and Body; trad.it.*Il canto degli antenati. Le origini della musica, del linguaggio, della mente e del corpo*, Torino, Codice.

Mongillo G. Machens K, Romo, Brody C.D. (2005), "Flexible control of mutual inhibition: a neural model of two interval discrimination", *Science* 307:1121:4).

Mongillo G., Barak O., Tsodyks M. (2008), "Synaptic theory of Working Memory", *Science* 319: 1543-46).

Montani, P. (2010), *L'immaginazione intermediale. Perlustrare, rifigurare, testimoniare il mondo visibile*, Roma–Bari, Laterza.

Morganti F., Riva G. (2006), *Conoscenza, comunicazione e tecnologia. Aspetti cognitivi della realtà virtuale*, Milano, edizioni Led.

Moscovici, S. (1993), *The Invention of Society*, Oxford, Polity Press.

Muscat, R. Shranz J.J. (1994), *Performer as Act of Faith*, GHE, Malta, 2004; *Id. Alla ricerca dell'uomo non progettato*, in "Teatro e storia", XVIII, n.25.

Niedenthal, P.M., Barsalou L.W. Winkielman, P. Krauth–Gruber S., Ric, F. (2005), *Embodiment in Attitudes, Social Perception, and Emotion*, in "Personality and Social Psychology Review", 9 (3), pp.184–111.

Oberman L.M., Pineda J.A., Ramachandran W.S. (2007), "The Human Mirror Neuron System: A Link between Action Observation and Social Skills", Soc. Cogn. Affect Neurosciences, 2 (1).

O'Keefe, J. e Dostrovsky, J. (1971), "The Hippocampus as a Spatial Map. Preliminary Evidence from Unit Activity in the Freely-Moving Rat", *Brain Research, 34*, 171-175.

Ong, W.J. (1982), *Orality and Literacy.The Technologizing of the Word* , London and New York, Methuen, 1982, trad.it.. it. *Oralità e scrittura*, Bologna, Il Mulino.

Panofski, E. (1973), *La prospettiva come «forma simbolica»,* Milano, Feltrinelli.

—. (1999), *Il significato nelle arti visive*, Torino, Einaudi.

—. (2009), *Rinascimento e rinascenze nell'arte occidentale*, Milano, Feltrinelli.

Perniola M. (1990), *Virtualità e perfezione*, in *Enigmi. Il momento egizio nella società e nell'arte,* Genova, Costa Nolan.

Plessner, H. (2006), *I gradi dell'organico e l'uomo*, Torino, Bollati Boringhieri. Rabin, J.F. "Repression', in: *International Dictionary of Psychoanalysis*, http://www.enotes.com/psychoanalysis-encyclopedia/repression.

Pradier, J.M. (1990), "La Scène et la fabrique des corps », in *Id. Towards a Biological Theory of the Body in Performance, New Theatre Quarterly*, Cambridge University Press, VI, 21, 1990: 86-98.

—. (2007), *Théatre et Neurosciences. Ou l'interminable aporie* in *Teatro e neuroscienze* (a cura di F. Bortoletti), , n.16: 73-87.

Prigogine, I. and Stengers I. (1979) *La nouvelle alliance, métamorphose de la science.*Paris/New-York: Gallimard/Bantam; trad.it. *La nuova allean-za. Metamorfosi della scienza, Torino, Einaudi.*

Ricciardi, M. (2010), *La comunicazione. Maestri e paradigmi*, Roma–Bari, Laterza.

Ricoeur, P. (2004), *Memory, History, Forgetting*, translated by K. Blamey and D. Fellauer. Chicago-London: The University of Chicago Press.

—. (2008), *Time and Narrative*, vol. 3, translated by K. Blamey and D.Fellauer. Chicago-London: The University of Chicago Press.

Riva G.A. (ed. by) (2003), *Ambient Intelligence The Evolution of Technology, Communication and Cognition Towards the Future of Human-computer interaction*, www.neurovr.org/emerging/volume.html

Rizzolatti, G., Fadiga, L., Gallese, V. Fogassi, L. (1996), "Premotor Cortex and the Recognition of the Motor action", in Cognitive Brain Research, 3, 1996, pp. 131–141.

Rizzolatti, G., Sinigaglia C. (2006), *So quel che fai. Il cervello che agisce e i neuroni–specchio*, Milano, Raffaello Cortina.

—. (2010), "The Functional Role of the Parietofrontal Mirror Circuit: Interporetations and Misinterpretations". *Nature Review Neuroscience*, 11 (4) 264-274.

Rommeru, C. (1992), 'Statut du corps dans les différentes religions et les différentes civilisations' in: Labesse, J. (ed) *Analyses et réfle-xionssur le corps*. Paris: Editions Marketing.

Sartre, J.P. (1969), *Being and Nothingness*, New York, Washington Square Press.

Schacter, L. (2007), *Alla ricerca della memoria. Il cervello, la mente e il passato*. Torino: Einaudi.

Schank, R.C. & Cleary. C. (1995), *Engines for Education*. Hillsdale, NJ:Erlbaum.

Schutz A., Luckman Th. (1973), *The Structures of Life-World*, Heinemann, London.

Sennet, R. (1978), *The Fall of Public Man*, New York, Random House.

Sills, D.L. (Ed.) (1968), *International Encyclopedia of the Social Sciences*, MacMillan, New York.

Simon, H. A. (1973), *Technology and Environment*, in "Management Science", XIX: 10: 110-1121.

Sofia G., (a cura di) (2009) *Dialoghi tra teatro e neuroscienze*, Roma, Alegre.

Symon, G., Long, K., Ellis, J. (1996), *The Coordination of Work Activities: Cooperation and Conflict in a Hospital Context*, in "Computer Supported Cooperative Work", 5, pp. 1–31.

Singer, C., Holmyard E.J., Hall, A.R, Trevor, W. A. (1954-1958), *History of Technology* (5 Voll.), Oxford University Press, New York.

Stern, D. N.(2005), *Il Momento presente in psicoterapia e nella vita quotidiana,* Milano, Raffaello Cortina.

Suchman, L.(2006), *Human–machine reconfigurations: Plans and Situated Action,* Cambridge, Cambridge University Press.

Suler, J. (2002), *Presence in Cyberspace*, in: http://www-usr.rider.edu

—. (2010), *The Basic Psychological Features of Cyberspace: Elements of aCyberpsychology Model.* http://usesr.rider.edu/~suler/psycyber/basicfeat.html

—. *Identity Management in Cyberspace,* htttp://users.rider.edu/~suler/psy-cyber/identitymanage.htm

Taylor, L.M. (2005), *Introducing Cognitive Development*, Hove, UK:Psychology Press.

Thaut, M. (2008), *Rhythm, Music, and the Brain: Scientific Èand Clinical Applications.* New York & London: Routledge.

Thaut, M., & Wheeler, B. (2010), "Music Therapy". In Juslin, P. and Sloboda, J. (Eds.) *Handbook of Music and Emotion: Theory, Research, Appli-cations*. Oxford: Oxford University Press.

Trevarthen, C. (2008), Foreword. Shared minds and the science of fiction: Why theories will differ. In Zlatev, J. et al. (Eds.) *The Shared Mind: Perspectives on Intersubjectivity,* Amsterdam, Philadelphia: John Benjamins Publishing Company.

Turkle, S. (2005), *The Second Self: Computer and the Human Spirit*, Cambridge (Mass.), The MIT Press.

Uexküll, J. von. (1934), *Streifzüge durch Umwelten von Tieren and Menschen,* Hamburg, Rowolt Taschenbuch Verlag; trad.it.

—. (1936), *I mondi invisibili,* Milano, Mondadori.

—. (1983), *Biological Rhythms and Human Adaptation to the Environment.*

Vallortigara, G. (1999), *La visione negli animali.* In Purghé, F., Stucchi, N.

—. (2000), *Lo studio comparato della cognizione animale,* Bologna, Il Mulino.

—. (2007), *Cervello di gallina. Visite (guidate) tra etologia e neuroscienze*, Torino, Bollati Boringhieri.

Varela, F.J., Thompson, E., Rosh, E. (1991), *The Embodied Mind: Cognitive Science and Human Experience*, Cambridge (MA), MIT Press

Vygotskij, L.S. (1978), *Mind in Society: the Development of Higher Psychological Processes,* Cambridge, MA, Harvard University Press; trad.it. 1980 *Il processo cognitivo*, Torino, Bollati Boringhieri.

Warf B., Arias S. (2009), *Spatial Turn. Interdisciplinary Perspectives*, New York, Routledge.

Young, T. Whitty, M. (2012), *Trascending Taboos.A Moral and Psychological Examination of Cyberspace;* London, Routledge.

Zerubavel, E. (2003), *Time Maps: Collective Memory and the Shape of the Past.* Chicago-London: The Chicago University Press.